MW01253352

Environment, Globalization and Economic Development

Environment, Globalization and Economic Development

Edited by

Dr. G. Rajaiah
Professor, Department of Economics,
Kakatiya University, Warangal

and

Dr. G. Bhaskar
Professor and Head, Department of Economics,
Kakatiya University, Warangal

New Century Publications
New Delhi, India

NEW CENTURY PUBLICATIONS
4800/24, Bharat Ram Road,
Ansari Road, Daryaganj,
New Delhi – 110 002 (India)

Tel.: 011-2324 7798, 4358 7398, 6539 6605
Fax: 011-4101 7798
E-mail: indiatax@vsnl.com • info@newcenturypublications.com
www.newcenturypublications.com

Editorial office:
LG–7, Aakarshan Bhawan,
4754-57/23, Ansari Road, Daryaganj,
New Delhi – 110 002

Tel.: 011-4356 0919

First Published: **2013**

ISBN: **978-81-7708-330-9**

Published by New Century Publications and printed at Salasar Imaging Systems, New Delhi.

Designs: Patch Creative Unit, New Delhi.

PRINTED IN INDIA

About the Book

Environmental pollution is one of the major problems faced by the world community, especially in the cities of the developing countries which have experienced unbridled growth of population, urbanization and industrialization.

The concept of sustainable development is inextricably linked with environment protection. It is a strategy for improving the quality of human life while living within the carrying capacity of supporting ecosystem. Sustainable development entails commitment, protection and preservation of environment.

Protection of the environment has to be a central part of any sustainable inclusive growth strategy. This aspect of development is especially important when consciousness of the dangers of environmental degradation has increased greatly.

This volume contains 12 research papers, authored by experts in the field, which provide deep insights into the relationship between economic development and environment in the context of increasing globalization of the world economies.

Editors' Profile

Dr. G. Rajaiah is presently Professor of Economics and Co-ordinator, SAP-DRS-II (UGC), Department of Economics, Kakatiya University, Warangal, Andhra Pradesh. He did his M.Phil. from Delhi School of Economics (DSE), University of Delhi and Ph.D. from Osmania University, Hyderabad. He was Principal, University Post-graduate College, Kakatiya University, Warangal. Apart from having 6 books to his credit, he has participated and presented papers in various conferences and seminars at the national and international levels.

Dr. Rajaiah is Vice-Chairman of Andhra Pradesh Economic Association. His areas of specialization include macroeconomics, development economics and environment.

Dr. G. Bhaskar is Professor and Head, Department of Economics, Kakatiya University, Warangal, Andhra Pradesh. He obtained M.Phil. and Ph.D. degress from Kakatiya University, Warangal. He has completed two minor Research Projects financed by University Grants Commission (UGC), New Delhi and published articles in professional journals of repute. His areas of specialization include agricultural economics, rural development and environmental problems.

Contents

Contributors

G. Bhaskar Professor and Head, Department of Economics, Kakatiya University, Warangal.

M. Yadagira Charyulu Reader in Economics, Pingle Government College for Women, Warangal.

M.N. Swarna Research Scholar, Department of Economics, Acharya Nagarjuna University, Guntur.

M. Rajani Research Scholar, Department of Economics, Acharya Nagarjuna University, Guntur.

B.S. Rao Professor, Department of Economics, Acharya Nagarjuna University, Guntur.

P. Tara Kumari Professor, Department of Economics, Andhra University, Visakhapatnam.

K. Siva Prasad Professor, Department of Economics, Andhra University, Visakhapatnam.

S. Vijay Kumar Reader in Economics, Kakatiya Government College, Hanamkonda.

M. Yadagiri Professor of Commerce, Telangana University, Nizamabad.

N. Sreenivas Lecturer in Business Management, Thushara Post-graduate School of Information Science and Technology, Rampur, Warangal.

K. Hari Babu Research Scholar, Department of Political Science and Public Administration, Acharya Nagarjuna University, Guntur.

K. V. R. Srinivas Research Scholar, Department of Political Science and Public Administration, Acharya Nagarjuna University, Guntur.

A. Srivasacharyulu	Research Scholar, Department of Political Science and Public Administration, Acharya Nagarjuna University, Guntur.
M. Galaiah	Research Scholar, Department of Political Science and Public Administration, Acharya Nagarjuna University, Guntur.
N. Alivelu Manga	Department of Economics, Kakatiya University, Warangal.
G. Rajaiah	Professor, Department of Economics, Kakatiya University, Warangal.
C. Basavaiah	Associate Professor and Head, Department of Economics, S.V.U.P.G. Centre, Kavali.
I. Malyadri	Assistant Professor, Department of Humanities and Sciences, P.B.R.V.I.T.S., Kavali.
J.V. Siva Kumar	Assistant Professor and Head, Department of Law, Telangana University, Nizamabad.
Dhulasi Birundha Varadarajan	Chairperson, Senior Professor and Research Scholar, Department of Environmental Economics, School of Economics, Madurai Kamaraj University, Madurai.

Preface

Environmental pollution can be defined as an undesirable change in the physical, chemical, or biological characteristics of the air, water or land that can harmfully affect health, survival or activities of human beings or other living organisms. Pollution refers essentially to a process by which a resource is rendered unfit for some beneficial use. Of the various kinds of pollution (air, water, land, noise, radiation and odour) that affect the quality of life in India, water pollution is by far the most serious in its implications for the health and well-being of the people.

Pollution has become a major threat to the very existence of mankind on this earth. Pollution of various resources has gone to such an extent that people are unable to breathe fresh air and drink fresh water. True, advancements in science and technology have added to human comforts by way of automobiles, electrical appliances, supersonic jets, modern medicines, communications and entertainment. However, these blessings have been diluted by serious problems posed by ecological imbalances and climate changes.

Sustainable development is that which meets the needs of the present without compromising the ability of future generations. The heaviest burden in international economic adjustment has been carried by the world's poorest population. The consequence has been the considerable increase in human distress and over exploitation of land and water resources to ensure survival in the short term. Further, the desire to solve all social and human problems at one stroke has influenced the pace of development. Its reaction is visible in the atmosphere and in natural, physical and social environment. Awareness regarding the consequences of this event has developed into a consensus referred to as ecological concern.

The concept of sustainable development is inextricably linked with environment protection. It is a strategy for improving the quality of human life while living within the carrying capacity of supporting eco system. It is clear from the above definition that sustainable development entails commitment, protection and preservation.

Environmental problems in India can be classified into two broad categories: (a) those arising from conditions of poverty and under-development and (b) those arising as negative effects of the very process of development. The first category has to do with the impact on the health and integrity of natural resources (land, soil, water, forests, wildlife etc.) as a result of poverty and the inadequate availability, for a large section of population, of the means to fulfil basic human needs (food, fuel, shelter, employment etc.). The second category pertains to unintended side effects of efforts to achieve rapid economic growth and development. In this category would fall the distortions imposed on national resources from poorly-planned development projects and programmes, and lack of attention to long-term concerns by commercial and vested interests. Thus, it is clear that concern for environment is essentially a desire for national development along rational sustainable lines. Environmental conservation is, in fact, the very basis of all development.

Environmental degradation in India has been caused by a variety of social, economic, institutional and technological factors. Rapidly growing population, urbanisation, industrial activities and increasing use of pesticides and fossil fuels have all resulted in considerable deterioration in the quality and sustainability of the environment.

There are considerable environmental risks facing the Indian economy due to high population density and the economic dependence of more than half the labour force on the natural resource base. These risks need to be identified and call for immediate preventive and remedial attention.

The risks are of three types: (a) growth-related risks resulting from uncontrolled release of industrial pollutants into the air and into water bodies, (b) poverty-related risks resulting from inadequate access to potable water, absence of adequate sanitation and indoor air pollution from burning freely collected biomass for cooking and (c) policy-induced environmental risks, several of which fall within the decision-making sphere of State Governments.

Protection of the environment has to be a central part of any sustainable inclusive growth strategy. This aspect of development is especially important when consciousness of the dangers of environmental degradation has increased greatly. Population growth, urbanization and anthropogenic development employing energy intensive technologies have resulted in injecting heavy loads of pollutants in the environment. More recently, the issue assumed special importance because of the accumulation of evidence of global warming and the associated climate change that it is likely to bring.

We place on record my deep appreciation and gratitude to all the contributors to this volume for their scholarly contributions and co-operation.

Warangal
2012

Dr. G. Rajaiah
Dr. G. Bhaskar

1

Environment-Development Interface

G. Bhaskar and M. Yadagira Charyulu

Introduction

Environmental pollution is one of the major problems faced by the world community, especially in the cities of the developing countries which have experienced unbridled growth of population, urbanization and industrialization. Municipal services—such as water supply and sanitation, drainage of storm water, management of solid and hazardous wastes, supply of adequate and safe food and housing—are unable to keep pace with urban growth. The unplanned location of industries in and around urban areas and volume of traffic has caused serious pollution problems. All these factors have led to deteriorating environmental conditions, adversely impacting the health of the people.

Natural and environmental resources are the basis of all economic activities. Soil, forests, mines, water, air and other natural resources are productive assets of an economy. In fact, productivity of an economic system depends, to a large extent, on the supply and quality of its natural resources. Poor quality of water and air adversely affects the health of population, causing reduced labour productivity and premature mortality. Similarly, soil degradation and erosion leads to reduced and inferior quality of agricultural output.

Economic activities, in turn, affect the quantity and quality of natural and environmental resources. Mining, lumbering, manufacturing, fishing and a host of other economic activities change the stock of natural resources which calls for appropriate trade-off between the needs of present and future generations.

Increase in economic activities—due to growing population and/or enhanced consumption levels—often impacts environment negatively. Degradation of natural resources and losses in environmental quality adversely affect people at large. As a consequence of high level of economic activities, emissions, effluents and waste discharges exceed the carrying capacity of different resources. Economic growth without environmental considerations can cause serious damage to the quality of life of the present and future generations. It is in this context that economists distinguish between economic growth and economic development. Economic growth means sustained increase in per capita real income. Economic development is a much wider term. It includes economic growth and at the same time encompasses questions regarding patterns of production, distribution of national income, consumption behaviour of the people and concern for environment.

Pollution can be defined as an undesirable change in the physical, chemical, or biological characteristics of the air, water or land that can harmfully affect health, survival or activities of human beings or other living organisms. Pollution refers essentially to a process by which a resource is rendered unfit for some beneficial use. Of the various kinds of pollution (air, water, land, noise, radiation and odour) that affect the quality of life in India, water pollution is by far the most serious in its implications for the health and well-being of the people.

Pollution has become a major threat to the very existence of mankind on this earth. Pollution of various resources has gone to such an extent that people are unable to breathe fresh air and drink fresh water. True, advancements in science and technology have added to human comforts by way of automobiles, electrical appliances, supersonic jets, modern medicines, communications and entertainment. However, these blessings have been diluted by serious problems posed by ecological imbalances and climate changes.

Sustainable Development

Sustainable development is that which meets the needs of the present without compromising the ability of future generations. The heaviest burden in international economic adjustment has been carried by the world's poorest population. The consequence has been the considerable increase in human distress and over exploitation of land and water resources to ensure survival in the short term. Further, the desire to solve all social and human problems at one stroke has influenced the pace of development. Its reaction is visible in the atmosphere and in natural, physical and social environment. Awareness regarding the consequences of this event has developed into a consensus referred to as ecological concern.

The concept of sustainable development is inextricably linked with environment protection. It is a strategy for improving the quality of human life while living within the carrying capacity of supporting eco system. It is clear from the above definition that sustainable development entails commitment, protection and preservation.

Sustainable development advocates economic progress in an environmentally responsible manner. Sustainable development attempts to strike a balance between the requirements of economic development and the need for protection of the environment. It seeks to combine the elements of economic efficiency, inter-generational equity, social concerns and environmental protection. Although, the term sustainable development has many interpretations, it generally refers to non-declining human well-being over time.

Economic development often depends on the utilisation and conversion of natural resources into useful goods and services. This process of consumption and conversion uses energy, minerals and other natural resources such as air, water, land and bio-diversity. It also produces wastes which are released into environment with adverse impacts.

Environment has become an important issue in recent years. The Stockholm Conference on Human Environment held in 1972 and the Earth Summit held in Rio-De-Janeiro in June 1992 [United Nations Conference on Environment and Development (UNCED)] served as catalysts to focus the attention of mankind on the deteriorating quality of environment.

Demographic expansion, over exploitation of natural resources, mega development projects—with least concern for sustainability and conservation—have greatly affected the environment. The magnitude and alterations that humanity can make to environment, threaten to undermine the basic natural process that sustains biosphere. The process of economic development necessarily entails exploitations of natural resources. However, when such exploitation becomes indiscriminate and is undertaken without adequate environmental safeguards, the consequences can be disastrous.

Environmental Economics

Environmental economics is a branch of economics which deals with the inter-relationships between environment and development. Environmental economics examines the interface between economic agents and environment. Economic activities of human beings have a profound impact on natural environment. Hence, use/abuse of natural resources has raised many moral, legal and practical questions for present and future generations. There has been a gradual shift in many developed countries from command and control type of policy approach to environmental policy goals. Presently, many governments require cost-benefit analysis of policy options with regard to changes in environmental legislations by resetting environmental standards and introducing new policy instruments for environmental protection. Environmental economics has also made significant contributions to valuation techniques and design of new policy instruments for pollution control and management.

Green growth involves rethinking growth strategies with regard to their impact on environmental sustainability and the environmental resources available to poor and vulnerable groups. It is significant to note that many stimulus packages announced globally to combat recession incorporated a green component. International experience is that green growth promotes inclusivity. Further, the renewable energy sector is relatively labour-intensive, with the potential for generating more jobs than the oil and gas industries.

Environment must not be considered as just another sector of national development. It should form a crucial guiding dimension for plans and programmes in each sector. This becomes clear only if the concern for environmental protection is understood in its proper context. Ecological degradation compromises the quality of life in the long-run.

United Nations Environment Programme (UNEP)

UNEP, established in 1972 after the Rio Earth Summit, is the principal entity within the UN System to assist the developing countries in building scientific and technical capacity, fostering partnership and knowledge development to promote environment for sustainable development. Based in Nairobi (Kenya), UNEP activities range from assessment of environmental trends—especially early warning systems to deal with environmental disasters and emergencies—to the promotion of environmental science and information.

One of the main responsibilities of UNEP is to keep under review the world environmental situation and ensure that emerging environmental problems of wide international significance are prioritized and receive appropriate and adequate consideration by the governments.

International Union for Conservation of Nature (IUCN)

IUCN is a unique global organization which started working in 1948. It is one of the handfuls of international organizations where governments and non-governmental bodies work together as parties. By virtue of being a member of IUCN, a country has access to the largest network of specialists in the field of conservation. IUCN is having an observer status at the United Nations and advises governments on matters related to conservation, integrity and diversity of nature and also ensures that any use of natural resources is equitable and ecologically sustainable.

India has a long relationship with IUCN. The Government of India was the first country in South Asia to join IUCN as a state member in 1969. It was also the first country in the region to host the General Assembly of IUCN in 1969.

Environment and World Trade Organization (WTO)

With privatisation, liberalisation and globalisation of the Indian economy, environment and forest sectors are also undergoing signs of change. Also, with the looming dangers of global warming and climate change, environment has emerged as a matter of great concern both at the national and international levels. Environment and forests sectors are increasingly figuring as areas of interests in the bilateral, plurilateral and multilateral free trade agreements. India being a founder member of the World Trade Organisation (WTO) is governed by its basic binding principles and has been actively participating in such trade negotiations. The Doha Round of trade negotiations launched in November 2001 introduced negotiations in environmental goods and services. As a result, both the environmental goods and environmental services have emerged as areas of significance for India. Further, Para 31(iii) of the Doha Ministerial Declaration (DMD) enjoins upon the WTO members to reduce or eliminate tariffs on environmental goods and services.

India's Environmental Resources and Problems

Soon after achieving Independence from Britain in 1947, India embarked on a programme of planned economic development with the launching of First Five Year Plan (1951-56). Ever since, the emphasis has been on removal of poverty and unemployment and all possible instruments of economic policy have been used to achieve the twin-objectives. However, the euphoria for rapid economic development created enormous pressure on country's natural resources. Forests were subjected to rapid degradation due to growing demand for forest-based inputs for industrial use. Other natural resources, namely water, land, and air became dumping grounds for industrial wastes, often with toxic materials. In short, environmental issues were completely overlooked during the first two decades (1950-70) of economic development.

It was in the early 1970s that Government realised the need for environmental protection as an integral part of industrial policy. Unfortunately, considerable damage had already been done to India's natural resources by the time preventive measures were taken. Thus, land/soil degradation has taken place due to floods, water logging, salination, indiscriminate mining and faulty agricultural practices. Similarly, forest wealth has dwindled due to withdrawal of forest products (timber, fuel wood), overgrazing, forest fires and location of development projects in forest areas.

India is the second most populous and seventh largest country in the world. Geographically, it accounts for a meagre 2.4 percent of the world's total surface area of 135.79 million square kilometres. Yet, India supports and sustains a whopping 16.7 percent of the world population. India covers an area of 32,87,263 square kilometres, extending from the snow-covered Himalayan peaks in the North to the tropical rain forests of the South. It has a land frontier of 15,200 kilometres. India's coast is 7,517 kilometres long of which 5,423 kilometres belongs to peninsular India, and 2,094 kilometres to the Andaman and Nicobar, and Lakshadweep Islands.

As regards its biological diversity, over 45,000 plant species are found in the country. The vascular flora which form the conspicuous vegetation cover itself comprises about 15,000 species. Several thousands of them are endemic to this country and they have so far not been reported from anywhere in the world. The biological diversity of the country is so rich that it may play a very important and crucial role in future for the survival of entire mankind, if it is conserved with utmost care.

India—with a varied terrain, topography, land use, geographic and climatic factors—has been divided into 10 recognizable bio-geographic zones and these zones together consist of 25 bio-geographic provinces. These zones encompass a variety of ecosystems: mountains, plateaus, rivers, forests, deserts, wetlands, lakes, mangroves, coral reefs, coasts and islands. India's diverse economy encompasses traditional village farming, modern agriculture, fisheries, handicrafts, a wide range of modern industries, and a multitude of services.

Environmental problems in India can be classified into two broad categories: (a) those arising from conditions of poverty and under-development and (b) those arising as negative effects of the very process of development. The first category has to do with the impact on the health and integrity of natural resources (land, soil, water, forests, wildlife etc.) as a result of poverty and the inadequate availability, for a large section of population, of the means to fulfil basic human needs (food, fuel, shelter, employment etc.). The second category pertains to unintended side effects of efforts to achieve rapid economic growth and development. In this category would fall the distortions imposed on national resources from poorly-planned development projects and programmes, and lack of attention to long-term concerns by commercial and vested interests. Thus, it is clear that concern for environment is essentially a desire for national development along rational sustainable lines. Environmental conservation is, in fact, the very basis of all development.

The ambient air quality of major cities and towns in the country is being monitored under the National Ambient Air Quality Monitoring (NAAQM) programme. The monitoring results indicate that sulphur dioxide and nitrogen dioxide levels are mostly within the permissible limits whereas Suspended Particulate Matter (SPM) values are higher than the prescribed limits at some places due to vehicular and air pollution, burning of fossil fuel and natural dusty conditions. The pollution caused by vehicles contributes significantly towards air pollution.

Noise levels in urban areas of India are excessively high due to rise in population, automobiles and industrial activities.

Similarly, the data on ambient water quality of rivers indicate that levels of coliform count and Biological Oxygen Demand (BOD) are generally high resulting in deterioration of water quality. This is primarily due to inadequate sanitation facilities and discharge of waste water into the surface water bodies without proper treatment. The dangers of environmental pollution are well-known. Thus, mercury poisoning results in neurological afflictions and exposure to toxic materials causes acute illness or even death.

The damage done to India's environment has been enormous. Ecological damage is often irreversible. Cleaning up later is not an option when terrestrial and aquatic biodiversity has been lost because of habitat destruction. For example, pollution and destructive fishing techniques have damaged a large proportion of coral reefs in some areas. As up to one-fourth of all marine species and one-fifth of known marine fish species live in coral reef ecosystems, the loss of reef habitats disproportionately threatens a high percentage of the ocean's plant and animal life. Complete reversal of this damage is unlikely. Efforts, therefore, need to focus on preserving biological resources before they are further damaged.

Causes of Environmental Degradation in India

Environmental degradation in India has been caused by a variety of social, economic, institutional and technological factors. Rapidly growing population, urbanisation, industrial activities and increasing use of pesticides and fossil fuels have all resulted in considerable deterioration in the quality and sustainability of the environment.

Population Pressure: India's growing population has aggravated the problems of poverty and unemployment. Poverty is said to be both cause and effect of environmental degradation. The circular link between poverty and environment is an extremely complex phenomenon. Inequality may foster unsustainability because the poor—who rely on natural resources more than the rich—deplete natural resources faster as they have no real prospects of gaining access to other types of resources. Moreover, degraded environment can accelerate the process of impoverishment, again because the poor depend directly on natural assets. Similarly, lack of employment opportunities in rural areas are driving the rural poor to metropolitan cities, creating urban slums and the problems associated with them.

The development model that has been followed in India so far does not meet the parameters of sustainable development. It is obvious from the present precarious state of environment in the country which is the result of uncontrolled urbanization and industrialisation with utter disregard for environmental concerns. The exponentially increasing population, relentless denudation of forest, falling of ground water table, increase in air and water pollution, are eroding the life sustaining resource base and posing serious threats to life, health and livelihood of the people.

Over Use of Water: Fresh water is perhaps the most important of all natural resources affected by uncontrolled economic growth. Experts have estimated that 30 years from now approximately one-third of the world population will suffer from water crisis. A growing scarcity of fresh water is now a major impediment to food production, eco system, health, social stability and peace among nations. Each year millions of tonnes of foodgrains are grown by depleting the underground aquifers. In large part of India, people are facing scarcity of water for both drinking and irrigation purposes.

Large stretches of all major rivers, lakes and other water bodies have become contaminated. Even the Ganges, the holiest of the holy rivers, has not been spared from pollution. Approximately 1,660 million litres of domestic and industrial effluents are discharged into the river every day.

Over exploitation of groundwater is also emerging as an increasingly serous problem in many parts of the country. In many districts of Tamil Nadu, Karnataka, Haryana, Rajasthan, Punjab and Gujarat, the groundwater table is declining at a rate of about half a metre to one metre per year. It is estimated that by the year 2017 there will be water crisis as per capita water availability will go down to 1,660 cubic metres which means water stress according to norms approved. Depleting forest cover (particularly in catchments areas), neglect of traditional water conservation techniques, increased pollution of both surface and ground water, improper water resource management, absence of sewage treatment plants and non-implementation of environmental laws in industries are the main factors contributing to water crisis.

Expansion of Urbanisation: Growing population leads to a greater concentration of people in the living areas. People move to urban areas abandoning rural settings in search of employment, comfort and facilities. As a result, cities are being over loaded with population that they can barely hold or support. A thickly populated area is the home of large number of vehicles, reservoir for solid and liquid wastes with poor sanitary conditions and management problems. In urban areas, the disposal of sewage and household wastes renders the water resources dirty and contaminated.

Rapid Industrialisation: Human needs are never ending. Discovery of new products and production of luxuries to suit the changing life styles are accompanied by the process of industrialisation. Industries, during the process of manufacturing of intermediate chemicals and products, generate waste materials and useless by-products as well. Each industry is associated with an emission of one type or other of pollutants or potential pollutants directly or indirectly. Not only are many industries responsible for pollution of air but also for the contamination of water. The quantity of water spent in producing every little thing in the world is unassumable large. Every process on the earth needs water, the universal solvent.

Industrial pollution is concentrated in industries like petroleum refineries, textiles, pulp and paper, industrial chemicals, iron and steel and non-metallic mineral products. Small-scale industries, especially foundries, chemical manufacturing and brick making, can also be significant polluters. In the power sector, thermal power, which constitutes bulk of the installed capacity for electricity generation, is an important source of air pollution. Hence, increased and efficient environmental vigilance is an absolute must for containing the negative environmental impact of industrialisation.

Adverse Impacts of Environmental Degradation

Environmental impacts resulting from the execution of development projects—like those involving thermal or hydro-power generation, mining industry, agriculture, human settlements etc.—manifest themselves through one or more of the environmental problems. For instance, mining operations in India have often led to serious problems of water and air pollution, land subsidence and scarring of large tracts of land. Indiscriminate discharge of wastes by industries has caused a whole variety of pollution problems including those due to heavy metals and other chemicals that are inimical to all life forms. The unplanned, intensive use of agricultural chemicals has led to cases of water pollution and appearance of pesticide residues in food and food products.

There are other serious and more insidious consequences for human health arising through poorly planned developmental activities. In particular, there is the whole range of tropical, communicable diseases such as malaria, filariasis, dengue, guinea worm, Japanese encephalitis etc. that are becoming more widespread due to the creation of favourable environmental conditions for the pathogens. The majority of the breeding places of these disease-vectors are created by man in the form of stagnant ponds, burrow pits and ditches.

While it is recognised that provision of water for agriculture and for human use is a major developmental activity and a vital necessity, it must be ensured that conditions favourable for the breeding of vectors of human and animal diseases are not encouraged. Unplanned urbanisation has changed the ecological conditions in favour of the spread of filariasis. Studies have shown that the construction of large reservoirs can result in the elevation of sub-soil water in the vicinity with consequent changes in the levels of fluoride, calcium, trace metals etc. in soil sediments. This in turn results in the re-emergence of diseases such as neurosis, in people who are forced to use the contaminated water. Skin infections, trachoma, guinea worm and schisto-somiasis are other diseases transmitted by water. The price for the lack of recognition and control of these environment related diseases is paid not only in terms of human health but also in terms of costs of pest control and medical care.

It is now recognised that most of these impacts can be minimised or even completely avoided by adequate pre-planning through the use of techniques like environmental impact analysis for which the inter-disciplinary expertise will need to be built up. Environmental considerations must form an integral part of all planning for development and be supplemented by mechanisms to ensure that environmental safeguard proposals are implemented and that there is systematic monitoring to assess their effectiveness.

It is the successful control of population growth and the satisfaction of basic human needs that will ultimately protect environmental health and hence the quality of life of the people. In that sense the entire plan for national development could be termed *environmental*. However specific programmes for environmental protection would also be necessary to correct various local and regional stresses on environmental resources arising as a result of the conditions of poverty and underdevelopment and the unintended side-effects of programmes for national development.

Environmental Risks Facing India

There are considerable environmental risks facing the Indian economy due to high population density and the economic dependence of more than half the labour force on the natural resource base. These risks need to be identified and call for immediate preventive and remedial attention.

The risks are of three types.

1. Growth-related risks resulting from uncontrolled release of industrial pollutants into the air and into water bodies.
2. Poverty-related risks resulting from inadequate access to potable water, absence of adequate sanitation and indoor air pollution from burning freely collected biomass for cooking.
3. Policy-induced environmental risks, several of which fall within the decision-making sphere of State Governments.

The widespread practice in States of zero-pricing electricity for farmers, has resulted in an alarming fall in ground water levels in many zones in the country, accompanied by soil salinity due to the conjunction of over-application of under-priced groundwater and poor drainage. In many States, surface irrigation water has a crop-specific rate structure, which is not crop-neutral and frequently carries an adverse incentive in terms of encouraging cultivation of water-intensive crops, even in regions that are water scarce. There are also policies of the Government of India which have added to the environmental risks facing the country.

Perhaps the most glaring example is the national fertiliser subsidy scheme. Uneven price interventions across nutrients have led to a decline in soil quality due to application of a distorted nutrient mix.

Summing Up

Protection of the environment has to be a central part of any sustainable inclusive growth strategy. This aspect of development is especially important when consciousness of the dangers of environmental degradation has increased greatly. Population growth, urbanization and anthropogenic development employing energy intensive technologies have resulted in injecting heavy loads of pollutants in the environment. More recently, the issue assumed special importance because of the accumulation of evidence of global warming and the associated climate change that it is likely to bring.

An important feature of any environmental strategy is that environmental objectives require action in several areas, which typically lie in the purview of different Ministries. The Ministry of Environment and Forests (MoEF), Government of India, has the important role of monitoring the development process and its environmental impact in a perspective of sustainable development and to devise suitable regulatory structures to achieve the desired results. While this role is crucial, environmental objectives can only be achieved if environmental concerns are internalized in policy making in a large number of sectors. This would require sharing of responsibility at all levels of government and across sectors with respect to monitoring of pollution, enforcement of regulations and development of programmes for mitigation and abatement. Regulatory enforcement must also be combined with incentives, including market and fiscal mechanisms to encourage both industry and people in their day to day working lives to act in a manner responsive to environmental concerns. Sustainable use of natural resources also requires community participation with a responsible role assigned to the communities for conservation.

Securing the environment is critical for India's future generations and not just a matter of international commitment. A degraded environment reduces the quality of life for all citizens, but the impact is particularly pronounced on the poor and vulnerable groups, as it is they who suffer the most from degraded access to clean water, air and sanitation, as well as from climate shocks. It is for this reason that, despite the fact that India's per capita greenhouse gas emissions are much below the world average and far lower than the average of developed countries, it has pursued policies which complement efforts towards mitigation of climate change. It is, therefore, important to incentivise economic policies that promote measures for energy conservation, renewable energy, soil conservation, afforestation and more effective and affordable access to clean water at different levels of government. This would impact all levels of government, including local bodies, which face mounting challenges in delivering better access to clean water, better solid waste management and enhanced (but green) local infrastructure.

In order to protect human health from the impacts of pollutants, efforts are needed to co-relate the manifestations of various diseases with the environmental factors especially respiratory diseases and cardiovascular diseases contracted due to exposure to various pollutants.

2

Sources and Effects of Air Pollution

M.N. Swarna, M. Rajani and B.S. Rao

Introduction

Air is an important commodity for every kind of life and vegetation. A man can survive without food for a few weeks, without water for a few days but his ability to survive without air is only a few minutes. A normal man breaths 22000 times a day and takes about 35 Ibs of air per day–far more than the requirement of food or water per day. A minimum of five million litres of air passes per year through the respiratory epithelium into alveoli of normal adult. The common thing is that we all breathe and pollute air. Air pollution comes from a variety of different sources like factories, power plants, cars, buses, trucks, wildfires etc. Air pollution doesn't only affect air, but also threatens the health of humans, trees and animals and is destroying the ozone layer which protects us from dangerous ultra-violet radiation. Statistically speaking, every day average an individual breathes more than 3,000 gallons of air and since many people live in urban areas full of smog, this affects their health causing them a lot of health problems.

Air Pollution: Meaning and Scope

To give a comprehensive definition of air pollution, it is necessary that we must understand the terms "Environment" and "Pollution". 'Environment' is the sum of substances and forces surrounding an organism in such a way that it has got some relationship with that particular organism. In relation to man, the 'environment' constitutes air, land, water, flora, fauna etc. because these have got direct or indirect relationship with man in a complex system. In other words "Environment" is a multidimensional system of complex inter-relationship in a continuing state of change. By "Environment" we mean not only our immediate surrounding but also a variety of issues connected with human activity, productivity, basic living and its impact on natural resources such as land, water, atmosphere, forests dam, habitat, health, energy resources, wild life etc.

The term 'pollution' has been derived from a Latin word. "Pollution" means and implies "to soil" or to "defile". Pollution implies an undesirable change in physical, chemical or biological characteristics of air, land, water, etc. which is likely to affect human health, animal or plant life. Therefore, the contamination of environment with impurities making it unfit for intended use is known as 'pollution'. Pollution may also be defined as contamination of air, water or soil with undesirable foreign elements or heat. Heat, itself, is not called as pollutant but it is capable of altering the climate conditions, which in turn can affect the life of plants, man and animals.

The World Health Organization (WHO) has defined "air pollution" as "substances" put into air by the activities of mankind, in concentration, sufficient to cause harmful effect to the health, vegetation, property or to interfere with the enjoyment of property.

Air pollution is the contamination of the air by such substances as fuel exhaust and smoke. It can harm the health of plants and animals and damage buildings and other structures. According to the World Health Organization, about one-fifth of the world's people are exposed to hazardous levels of air pollutants. In scientific terms air pollution can be defined as the presence of one or more foreign constituents in natural composition of air. Air pollution occurs when wastes dirty the air. People produce most of the wastes that cause air pollution. Such wastes can be in the form of gases or particulates. These substances result chiefly from burning fuel to power motor vehicles and to heat buildings. Industrial processes and the burning of garbage also contribute to air pollution. Natural pollutants (impurities) include dust, pollen, soil particles, and naturally occurring gases.

The rapid growth of population and industry, and the increased use of automobiles and airplanes, has made air pollution a serious problem. The air we breathe has become so filled with pollutants that it can cause health problems. Polluted air also harms plants, animals, buildings materials, and fabrics. In addition, it causes damage by altering the earth's atmosphere. The damage caused by air pollution costs the people billions of dollars each year. This includes money spent for health care and increased maintenance of buildings. Air pollution may cause damage to the environment that cannot be reversed.

Composition of the Atmosphere

The present atmosphere of the Earth is probably not its original atmosphere. The original atmosphere may have been similar to the composition of the solar nebula and close to the present composition of the Gas Giant planets. The earlier atmosphere was lost to space and replaced by compounds out gassed from the crust or (in some more recent theories) much of the atmosphere may have come instead from the impacts of comets and other planetesimals rich in volatile materials.

As per the present estimate the composition of the air is 79 percent Nitrogen (N_2), 20 percent Oxygen (O_2), the remaining 1 percent is consists of several gases such as CO_2, Argon, Helium etc.

Layers of the Atmosphere

The atmosphere of the Earth may be divided into several distinct layers.

Troposphere

The troposphere is where all weather takes place; it is the region of rising and falling packets of air. The air pressure at the top of the troposphere is only 10 percent of that at sea level

(0.1 atmospheres). There is a thin buffer zone between the troposphere and the next layer called the tropopause.

Stratosphere and Ozone Layer

Above the troposphere is the stratosphere, where air flow is mostly horizontal. The thin ozone layer in the upper stratosphere has a high concentration of ozone, a particularly reactive form of oxygen. This layer is primarily responsible for absorbing the ultraviolet radiation from the Sun. The formation of this layer is a delicate matter, since only when oxygen is produced in the atmosphere can an ozone layer form and prevent an intense flux of ultraviolet radiation from reaching the surface, where it is quite hazardous to the evolution of life. There is considerable recent concern that manmade flourocarbon compounds may be depleting the ozone layer, with dire future consequences for life on the Earth.

Mesosphere and Ionosphere

Above the stratosphere is the mesosphere and above that is the ionosphere (or thermosphere), where many atoms are ionized (have gained or lost electrons so they have a net electrical charge). The ionosphere is very thin, but it is where aurora take place, and is also responsible for absorbing the most energetic photons from the Sun, and for reflecting radio waves, thereby making long-distance radio communication possible. The structure of the ionosphere is strongly influenced by the charged particle wind from the Sun (solar wind), which is in turn governed by the level of solar activity. One measure of the structure of the ionosphere is the free electron density, which is an indicator of the degree of ionization.

Sources of Air Pollution

The sources may be natural or anthropogenic (man-made). Natural sources include volcanic eruptions, forest fires, cosmic dust, pollen grains, sand storms, hydrogen sulphide and methane from anaerobic decomposition of organic matter, etc.

Man-made sources such as burning of fossil fuels, emissions from vehicles, rapid industrialization, agricultural activities, warfare's, etc. are major causes of air pollution. The sources may also be classified as stationary or mobile sources (Table 2.1).

Table 2.1: Sources of Air Pollutants

Stationary Sources	Point Sources	These are large stationary sources, such as, industries, power plants, municipal incinerators, etc.
	Area Sources	These are small stationary sources and mobile sources with indefinite routes, such as, residential heating, commercial and institutional heating, open burning, city traffic, etc.
Mobile Sources	Line Sources	These are highways, railway tracks, navigation routes, etc.
	Area Sources	These are airports, railway stations, ports, etc.

Classification of Air Pollutants

Air pollutants may be classified according to origin, chemical composition and state of matter.

According to Origin: On the basis of origin, air pollutants can be divided into two categories–primary and secondary air pollutants.

Primary air pollutants are those which are emitted directly to the atmosphere and found there in the form in which they are emitted. For example, particulates, carbon monoxide (CO), oxides of sulphur (SO_x), oxides of nitrogen (NO_x), hydrocarbons, etc. The five main primary air pollutants (viz. particulates, CO, SO_x, NO_x and HCs) contribute more than 90 percent of global air pollution.

Secondary air pollutants are those which are produced in the air by the interaction among two or more primary air pollutants, or by reaction with normal atmospheric constituents, with or without photo activation. For example, ozone (O_3), peroxyacetyl nitrates (PAN), formaldehyde, formation of acid mists, smog (coal induced and photochemical smog), etc.

According to Chemical Composition: On the basis of chemical composition, air pollutants can be divided as organic and inorganic air pollutants. Organic compounds contain carbon and hydrogen, and many also contain certain elements such as oxygen, nitrogen, sulphur and phosphorus. Examples of organic air pollutants are hydrocarbons, aldehydes ketones, carboxylic acids, organic sulphur compounds, etc. Inorganic air pollutants include compounds, such as CO, CO_2, SO_x, NO_3, etc.

According to State of Matter: On this basis, air pollutants are classified as particulate and gaseous air pollutants. Particulate air pollutants include finely divided solids and liquids dispersed in gaseous media. Dust, smoke, fly ash, flumes etc. are examples of solid particulates; while mist, spray, fog, etc., are liquid particulate air pollutants. Gaseous air pollutants are organic gases like benzene, methane, butane, aldehydes, ketones, etc., as well as inorganic gases like CO_2, SO_x, CO, NH_3, NO_x, etc.

Effects of Air Pollution

The gaseous and particulate materials added to the atmosphere by the activities of man are considered to be pollutant when their concentrations are sufficient to produce any harmful effects. The majority of man-made emissions to the atmosphere also have natural sources and in many cases these are larger than the pollutant ones.

After formation, pollutants are emitted to the atmosphere and dispersed. Once mixed with the air, some air pollutants such as the inert fluorinated hydrocarbons used in sprays– persist unaltered and become mixed throughout the atmosphere where they potentially have a global influence. More reactive pollutants have a shorter lifetime in the atmosphere and are removed either by conversion to normal atmospheric constituents or by deposition on the surface of the earth. In the process they may react with other atmospheric constituents to form secondary pollutants, which are also removed by the same process. Both the primary pollutants and the secondary pollutants can cause alteration to the chemical composition of soils and waters, and direct damage to biological systems and property. In certain cases, a synergistic interaction occurs where the total effect is enhanced over and above sum of the effects of the individual pollutants present.

Effects on Vegetation: Air pollution can affect plants to varying degrees. At the lowest levels, i.e. below the 'threshold', there is no effect, such as visible damage, cumulative chronic effects, genetic effects or even gradual changes in the composition of the plant community. However, event at these level air pollutants can be stored in the plants and introduced into the food chain, affecting animals which eat the plants.

The entry of air pollutants to plants may take place directly by gaseous diffusion or from the contaminated soil, acidic air pollutant in particular. The direct entry of gaseous air pollutants like SO_x, NO_x, and CO2 etc. may take place directly by stomata of the foliages. Solid particulates are, however, adsorbed on the surface. In general, pollutants cause injury at lethal concentration on sensitive plants while tolerant plant species are capable of sinking the pollutants to a considerable extent without any injury.

Thus, various air pollutants have different types of injuries on exposed sensitive plants. Suspended particulates after deposition on foliages cause a number of damages to leaf functions, viz.

1. Changes in the sun's energy falling on the leaf surface affecting the energy exchange due to dust layer;
2. Decreases chlorophyll content;
3. Interruption in gaseous exchange due to clogging of stomata by dust particulates; and
4. Dust deposition causes changes in the soil properties that

support the plant growth.

Effects on Animals: The indirect effect of air pollutants has been observed for a considerable time. It tends to occur near smelters treating non-ferrous ores, and near factories, such as phosphate fertilizer works, brick kilns and aluminum smelters, where fluorides are emitted and are concentrated in the grasses in surrounding fields. Identically lead compounds from automobile exhausts are deposited near to roads, although the concentrations found in vegetation are much smaller than those near smellers and have so far been measured as below the accepted threshold for toxicity to animals. But with time such metals accumulated in animal bodies through food chain and finally posing danger to animal health. The signs of heavy metal poisoning are diarrhoea, anaemia and stiffness. Fluorosis is a more widespread problem, which affects ruminants, particularly dairy cows.

Effects on Man: Air pollutant along with breathing air enters the nose, where fine hair filter out most particles greater than about 10 micrometers diameter. The air is then warmed and humidified, and is then passed through the windpipe into the bronchial tubes which subdivide the air stream and pass it into the lungs where there is a multiplicity of air sacs. From air sac through capillary diffusion gaseous pollutants enter into blood streams and particulates deposit in the alveolar sacs. The dangers of some small particles such as silica and asbestos, which are common in mines, quarries and some industrial plants, are well known over the years. They lead to specific occupational diseases such as silicosis or asbestosis.

In general, air pollutants mostly enter through the respiratory passages and thereby cause various types of health disorders. Limits are set as to the maximum concentrations to which healthy persons are permitted to be exposed for periods of up to 8 hours per day. These are called Threshold Limit Values (TLV).

Effect on Materials: Air pollutants have a delirious effect on materials; stone, paintwork, stained glass, fibre material and others. The soiling effect of particulate is obvious in industrial cities where building of light-coloured stones and bricks soon take on the characteristic black colour. The erosion of the stonework on buildings of great historic and architectural value is very serious indeed; other results of air pollution are the faster deterioration of clothing, curtains and wood, the corrosion of metals and the soiling and subsequent cracking of paintwork. The rapid deterioration of rubber acted on by ozone can be dioxide in industrial cities.

Global Climate Changes: Over past couple of decades, it was realized that air pollutants cause a considerable change in global climate and associated processes, viz., green house effects, ozone depletion, acid precipitation and Elnino effects etc.

Damage of Monuments by Air Pollutants

The pollutants of the atmosphere, viz., CO_2, SO_2 and NO_2 often produce acid after dissolution with atmospheric moisture. As such they produce bulk of carbonic acid and traces of sulphuric and nitric acid. Such acidic water may cause gradual corrosion of materials by which monuments were made of.

Primarily limestone blocks (chalk to marble) of the monuments were affected. Rainfall containing carbonic acid slowly attacks limestone to give a solution of calcium bicarbonate, which then transforms to calcium carbonate. The action of oxides of sulphur in polluted air in the presence of moisture is to convert the calcium and magnesium carbonates of stone to calcium and magnesium sulphates, calcium sulphate is slightly soluble in water and magnesium sulphate is very soluble. On drying the solutions, the calcium, and magnesium sulphates, form hydrated crystals of greater volume than the original carbonates, with the result that encrustations and friable scale are formed to produce effects known as blistering, flaking and exfoliation.

Slates containing carbonates and the calcareous sandstones, often used as roofing materials, are attacked by pollutants, and atmospheres. Damage usually occurs mainly on the undersides, especially between the laps where water is held for some time as a thin film. Though brick work can soon become badly discoloured by soot and grime, most types are fairly resistant to damage by sulphur oxides in the air, unless the bricks absorb sulphates produced by the action of the oxides of sulphur on adjacent mortar limestone.

Recent debate regarding the damage of internationally reputed monuments like Taj Mahal of Agra (India) and Victoria Memorial of Kolkata (India) are considered to be very significant in this regard.

Conclusion

A clean environment is the basic thing which is essential for the existence of life. Environmental pollution is a natural and essential concomitant of the activities of living organisms. Until recently, environmental pollution problems have been local and minimal because of the earth's own ability to absorb and purify minor quantities of pollutants and to maintain the ecological balance. Environmental pollution can be described as unfavourable alteration of our surrounds, through direct or indirect effects of changes in energy pattern, radiation levels, chemical and physical constitution and abundances or organisms. These changes may affect humans directly or through the supplies of water, air and agricultural other biological products and prove detrimental to their physical development, health, efficient performance, comforts and enjoyment of living today and the future.

However, air pollution is increasing with an alarming rate all over. Preventing pollution is a burning issue in the world today. To prevent air pollution controls the factors which are responsible for the problems of environmental degradation and air pollution is necessary. Such responsible factors are industrialization, urbanization, deforestation, population explosion, excessive use of pesticides in agriculture, poor sanitary and drainage system, increasing number of vehicles etc. Awareness of the public and public participation is also crucial for the prevention. Every citizen has a role to play in preventing pollution. We must take responsibility for our actions, weather as individuals or as members of a community or an organization. When citizens, governments and industry come together to build a partnership for the environment, they are taking big step toward ensuring sustainability.

3

Women Workers in the Visakhapatnam Special Economic Zone (VSEZ)

P. Tara Kumari and K. Siva Prasad

Introduction

Export Processing Zones (EPZs) constitute an important element of promotion in developing countries. Besides developing physical infrastructure, they provide physical and taxation incentives to attract overseas investment. Boosting of exports, foreign exchange earnings, foreign investments, technology transfer, stimulating foreign direct investment, generating employment and developing forward and backward linkages have been the major objectives of EPZs. The EPZs work to the advantage of the both the investors and the host country. The host country is able to earn foreign exchange and generate employment and the foreign investors gain the competitive edge by drafting cheap labour and other inputs in the process of manufacture.

A unique feature of EPZ employment structure is its feminization due to the over whelming presence of electronics, gem and jewellery, textile and garment units. Assembling operations in these units are repetitive in nature and require low-level skills for which women are popularly considered to be more eminently suited. It has been observed that employment of women workers is high in the garment, gem and jewellery industry, next only to data processing and electronic assembling units. Employers have a strong preference for women workers since it is believed that dexterity, patience and other specific attributes make them more suitable than men for carrying out tasks that need painstaking attention. Women regardless of their level of education constitute the bulk of the unskilled and semiskilled workers and that is particularly pronounced in the textile, clothing, leather, knitwear, electronics, optical and diamond industries.

"Skill development programme involves learning by doing," workers are drawn mainly from the rural unskilled female category that learned their skills through work under a supervisor for a period of time. If, after the specific period, she cannot graduate to the permanent category she will again undergo supervised training for another term. Skill development through learning by doing continues at other stages of operation as well (Gupta, 1996). [1]

Though the EPZs are achieving their objective of promoting exports they failed in achieving the objective of improving the welfare of working environment of weaker sections especially the women. Therefore, they need to examine the employment conditions of women workers and how the women were exploited in the units established in EPZs in general and VSEZ in particular. To make appropriate policy intervention this study is and attempt in that direction.

This present study of women workers in VSEZ is a modest attempt to analyse toe problems faced by women workers with particular focus on the working conditions, welfare facilities, and working environment.

Data Base and Methodology

The present study is based on a sample survey in four units operating in Visakhapatnam Special Economic Zone in Visakhapatnam, covering a sample of 311 workers employed in these plants. The four units are World Wide Diamond (WWD), Madras Knitwear Pvt. Ltd. (MKW), Bangalore Swatters Pvt. Ltd. (BSW), and L.I.D. Jewellery Pvt. Ltd (LID). The random sample of 311 women was from these plants, which gave permission to collect information from their workers. After getting the initial information about the plants and the workers, a structured interview schedule consisting of both close-ended and open ended questions relating to working environment of women workers was canvassed. The collected data was coded and tabulated systematically.

Table 3.1: Gender-wise Distribution of Workers in the VSEZ Units

S. No.	Units	Male	Female	Total
1	World Wide Diamond	628 (69.16)	280 (30.84)	908
2	Madras Knitwear Pvt. Ltd.	103 (31.40)	225 (68.60)	328
3	Bangalore Swatter Pvt. Ltd.	55 (36.67)	95 (63.33)	150
4	L.I.D. Jewellery Pvt. Ltd.	409 (50.25)	405 (49.75)	814
Total		1195 (54.32)	1005 (45.68)	2200

Note: Figures in the parenthesis indicate percentages

The data in Table 3.1 indicates that the women workforce of the four units put together is accounting for about 45.68 percent. Among these four units WWD and LID shows that the proportion of female workforce is less than the male workforce. Where as female workforce is higher in other two units i.e. MDW and BSW.

Period of service is an important tool in measuring the workers hold on their employment. It indicates that the more the number of years of service, the more the security of employment. The distribution of workers according to their service with the percent employers is presented in Table 3.2. A majority of the workers i.e., 44.69 percent have put on three to four years of service with the present employers while 35.38 percent have below two years of service and 18.97 percent of workers have five to six years of service. Only a marginal percentage of (0.96) workers are in the service for above seven years.

Table 3.2: Working Experience of Women in the VSEZ Units

S. No.	Experience (*in years*)	VSEZ Units				
		WWD	MKW	BSW	LID	Total
1	<Two	12 (10.91)	21 (19.10)	16 (14.55)	61 (55.45)	110 (35.38)
2	Three-Four	56 (40.29)	26 (18.71)	25 (17.99)	32 (23.02)	139 (44.69)
3	Five-Six	35 (59.32)	05 (8.47)	09 (15.25)	10 (16.95)	59 (18.97)
4	>Seven	0	0	03 (100.00)	0	03 (0.96)
Total		103 (33.12)	52 (16.72)	53 (17.04)	103 (33.12)	311 (100.00)

Note: Figures in the parenthesis indicate percentages

Among the four units of the VSEZ working experience of women below two years is highest in LID (55.45 percent) and WWD (10.91 percent). Whereas among women having three to four years and five to six years of experience it is highest in WWD (40.29 percent) followed by LID (23.02 percent). On the whole it may be inferred from the Table that the women workers gained better experience than other units in WWD. Even then, those who gained more experience still may not question the principal employer to provide better facilities because of the fear of insecurity of employment.

Table 3.3: Training Period (in months) of Women Workers in the VSEZ Units

SI.No	Training Period (Months)	VSEZ Units				
		WWD	MKW	BSW	LID	Total
1	Six	0	52	53	0	105 (33.76)
2	Twelve	103	0	0	103	206 (66.24)
Total		103 (33.12)	52 (16.72)	53 (17.04)	103 (33.12)	311 (100.00)

The women workers come from rural areas where life is slow with no pressure on time. The workers adjust remarkably well to the new environment but some problems remain, e.g. absenteeism, lack of commitment and low productivity. In order to enhance the skills and ability and to chose and shape their occupations, training is essential for the development of women. In order to examine the importance of training in the units of VSEZ, Table 3.3 shows the training period provided by different units for women respondents. It is observed from the Table that two units are giving six months training i.e., MKW and BSW, another two units are giving one year training, i.e. WWD and LID. The proportion of the units giving six months training is nearly, 34 percent and the proportion of the units giving twelve months training is 66 percent.

Table 3.4: Leave Provisions for Women Workers in the VSEZ Units

SI.No.	Type of Leaves (Days)	VSEZ Units			
		WWD	MKW	BSW	LID
1	Casual Leave	12	10	10	15
2	Earned Leave	15	12	15	15
3	Sick Leave	10	12	12	10
4	Maternity Leave	90*	90**	90*	90*

Note: *With 75 percent salary; **with 50 percent salary

Table 3.4 shows that the leave provision for women workers in the units of VSEZ. The workers in these industries are eligible for twelve days of casual leave in some units. There are 10-15 days leave per year. And there is provision for earned leaves for 15 days in WWD, BSW, LID and 12 days in MKW per year. In VSEZ units, employees are eligible for 10-12 days of sick leave per year. There is provision for maternity leave for three months (90 days with 50 of 75 percent of salary). Even Sundays are working days for them. The provision of leave is based on rotation. No possibility of availing public holidays. If the workers avail leave without prior permission, salary will be deducted.

Durgan and Sinha did a study [2] on 'job satisfaction and the complex of various attitudes possessed by an employee. These attitudes are related to the many aspects of the jobs such as wages, supervision, steadiness of employment, conditions of work, opportunities for advancement and such other specific factors.

They concluded that the office group was more dissatisfied with their work.

Table 3.5: Job/Work Satisfaction of Women Workers in the VSEZ Units

S. No.	Rate of Satisfaction	VSEZ Units				
		WWD	MKW	BSW	LID	Total
1	Very much	30 (76.92)	02 (5.13)	01 (2.56)	06 (15.38)	39 (12.54)
2	Moderate	26 (38.81)	03 (4.48)	15 (22.39)	23 (34.33)	67 (21.54)
3	Like that	28 (31.82)	18 (20.45)	18 (20.45)	24 (27.27)	88. (28.30)
4	Not much	19 (16.24)	29 (24.79)	19 (16.24)	50 (42.74)	117 (37.62)
	Total	103 (33.12)	52 (16.72)	53 (17.04)	103 (33.12)	311 (100.00)

Note: Figures in the parenthesis indicate percentages

Table 3.5 presents distribution of respondents in relation to job satisfaction. When respondents were asked whether they are satisfied with their present job there emerged four categories of respondents. It is evident from the Table that one third of the respondents showed satisfaction with their present job and out of this 12.54 percent said that they like this work very much. About 37.62 percent workers expressed that the rate of satisfaction is 'not much'. On the whole, it is important to note that only 12.54 percent workers belong to the categories of' very much' rate of satisfaction. Thus, the job satisfaction in the present Study is considered to be a product of satisfaction with one's skills, wages, satisfactory relations with supervisors and physical working conditions.

Desai's study of worker's expectations from supervisors and management reflects the role of supervisors between the workers and the management. The study highlights the following characteristics for a good supervisor, viz., to treat them as human beings, to help them in their work, be friendly, cooperative, impartial and not using much authority. [3]

Table 3.6: Working Relationships with Management of Women Workers in the VSEZ

S. No.	Rate of Satisfaction	VSEZ Units				
		WWD	MKW	BSW	LID	Total
1	Very Cordial	10 (16.95)	16 (27.12)	15 (25.42)	18 (30.51)	59 (18.97)
2	Cordial	18 (38.30)	08 (17.02)	09 (19.15)	12 (25.53)	47 (15.11)
3	Partly Cordial	25 (27.47)	20 (21.98)	19 (20.88)	27 (29.67)	91 (29.26)
4	Not Cordial	50 (43.86)	08 (7.02)	10 (8.77)	46 (40.35)	114 (36.66)
Total		103 (33.12)	52 (16.72)	53 (17.04)	103 (33.12)	311 (100.00)

Note: Figures in the parenthesis indicate percentages

Satisfaction regarding relations with management is presented in Table 3.6. These relations with management are reported to be partly cordial and not cordial with respect to a majority of respondents (nearly 65.92 percent). Only 18.97 percent are considered to be very cordial and 15.11 percent have cordial relations with their management. Some of the women workers grieved about the misbehaviour of the certain personnel in the management. There are even cases of sexual exploitation in night shifts by the managers. Even though they are doing their best, some supervisors insist the women workers to do the work speedily.

Table 3.7: Suggestions to Improve the Quality of Employment Conditions and Working Environment by Women Workers Environment by Women Workers in the VSEZ Units

S.No.	Suggestions	VSEZ Units				
		WWD	MKW	BSW	LID	Total
1	Medical Facility with full salary	18 (32.14)	10 (17.86)	08 (14.29)	20 (35.71)	56 (18.01)
2	Maternity leave with full salary	15 (35.71)	09 (21.43)	06 (14.29)	12 (28.57)	42 (13.50)
3	Medical Check-up for every six months	10 (23.81)	07 (16.67)	07 (16.67)	18 (42.86)	42 (13.50)
4	Medical reimbursement for private treatment	15 (40.54)	05 (13.51)	04 (10.81)	13 (35.14)	37 (11.90)
5	50 percent reimbursement in emergency treatments	10 (33.33)	05 (16.67)	05 (16.67)	10 (33.33)	30 (9.65)
6	Crèche Facility	08 (40.00)	02 (10.00)	06 (30.00)	04 (20.00)	20 (6.43)
7	Company Hospital with Specialties	14 (34.15)	08 (19.51)	10 (24.39)	09 (21.95)	41 (13.18)
8	Company Bus facility	07 (22.58)	08 (19.35)	07 (22.58)	11 (35.48)	31 (9.97)
9	Uniformity in Salaries	06 (50.00)	0	0	06 (50.00)	12 (3.86)
Total		103 (33.12)	52 (16.72)	53 (17.04)	103 (33.12)	311 (100.00)

Note: Figures in the parenthesis indicate percentages

Table 3.7 gives the details of the suggestions given by the respondents to improve their quality of employment conditions and working environment. Majority of the respondents suggested that medical and maternity leave should be provided with full salary. And some of them expected to have free medical check-up in every six months. The respondents expressed to have a separate hospital on their own with all specializations. Apart from this they expected to have reimbursement for private treatment, 50 percent in case of emergencies, crèche facility, bus facility and uniformity in their salaries. Uniformity means equal payments for equal work to those who joined at the same time.

Conclusion

In the VSEZ, women constitute a considerable proportion of the labour force. The nature of employment reveals the fact that except a few, all the workers are temporary, which show the insecurity of employment. The jobs they do are unskilled or low skilled. Men are given preference in the skilled and high profit jobs. At policy level, the importance of the distinct economic role of women in the zone is not reflected. On papers the national labour laws are applicable in the zone but in practice things are different. The working conditions and work environment show long hours of work, standing posture of work, lack of job security, lack of personal protective equipment, absence of labour unions, low wage rates and absence of non-wage benefits.

Finally, working environmental conditions of the VSEZ women workers are very poor and unhealthy.

End Notes

1. S.P. Gupta (ed.) (1996): China's Economic reforms: The Role of Special Economic Zones and Economic and Technological Development Zones, Allied Publishers Ltd., New Delhi.
2. Durgananda Sinha (1969): "Job Satisfaction in Office and Manual Workers," The Indian Journal of Social Work, Vol. XIX, No. 1, June, pp. 39-46.
3. Desai, K.G. (1969): "A Study of Workers' Expectations from Supervisors and Management", The Indian Journal of Social Work, Vol. XXX, No. 2. July, pp. 105-115.

4

Environmental Concerns for Sustainable Development

S. Vijay Kumar

Economic development is a process whereby an economy's real national income increases over a long period of time. There is direct relationship between environment and economy. Economic development without environmental considerations can cause serious environmental damage in turn impairing the quality of life of present and future generations. In the process of economic development, the environmental problems have been ignored or less concentrated. Now, the need of the hour is to concentrate on sustainable development. Sustainable development means, "Meeting the needs of present generation without compromising with the needs of future generations."

In 1983, the United Nations set up the World Commission on Environment and Development called 'the Brundtland Commission' to examine the problems related to this area. The Commission in its report entitled "Our Common Future" submitted in 1987, used and defined this concept of sustainable development for the first time.

Objectives of Sustainable Development

Sustainable development aims at the creation of the sustainable improvements in the quality of life for all people and this should be the principal goal of development policy. Accordingly, the main objectives of sustainable development are stated as under:

(1) accelerating economic growth, (2) meeting basic needs, (3) raising the living standards, (4) helping in ensuring clean

environment free from all types of pollution, (5) maximizing the net effects of economic development, (6) preservation and enhancement of the stock of the environmental, human and physical capital, (7) inter-generational equity, and (8) overall strict control on gross exploitation of the natural resources of each country.

Environment-Economy Linkages

All economic activities either affect or are affected by natural and environmental resources. Activities such as extraction, processing, manufacture, transport, consumption and disposal change the stock of natural resources add stress to the environmental systems and introduce Wastes to environmental media. Moreover, economic activities today affect the stock of natural resources available for the welfare of the future and have intertemporal welfare effects. From this perspective, the productivity of an economic system depends in part on the supply and quality of natural and environmental resources.

Natural and environmental resources have three economic roles: (1) waste disposal services, (2) natural resources inputs into production, and (3) directly consumed life support services and authentic amenities. The natural and environmental resource input function is central to understanding the relationship between economic growth and environment. Water, soil, air, biological, forest and fisheries resources are productive assets, whose quality helps to determine the productivity of the economy.

Focusing on this, the role of environment as producer highlights the direct effect the environmental problems have on economic growth. Thus economic management lays emphasis on the environment and the environmental quality in turn on the efficient working of the economy. Environmental degradation imposes costs on the economy that results in output and human capital losses.

Loss of labour productivity resulting from ill health,

foregone crop output due to soil degradation and erosion, lost fisheries output and tourism receipts from coastal erosion or lost soil productivity from deforestation can be some of the manifestations of such reduced output. Moreover, growing bodies of epidemiological studies suggest that air and water population are taking a heavy toll, particularly of people in the developing world, through ill health and premature mortality. The impact of water and air pollution is particularly adverse on the younger, the very aged and the poor. Pollution control is thus linked to sustainable development and not a luxury good to be afforded after the development process has taken off.

Major Environmental Concerns

A country's environmental problems vary with its stage of development; structure of its economy, production technologies in use and its environmental policies. While some problems may be associated with the lack of economic development, others are exacerbated by the growth of economic activity. Poverty presents special problems for a densely populated country with limited resources.

Soil Degradation: In India, out of the total geographical areas of 329 million hectares, 175 million hectares is considered degraded. Erosion by water and wind is the most important factor to soil erosion with other factors like water logging, salivation etc., adding to the degradation. While soil erosion by rain and river in hill areas causes landslides and floods, deforestation, overgrazing, traditional agricultural practices, mining and incorrect setting of development projects in forest areas have resulted in opening up of these areas to heavy soil erosion. In the arid west, wind erosion causes expansion of desert, dust storms, whirlwinds and destruction of crops, while moving sand covers the land and makes it sterile. In the plains, riverine erosion due to floods and eutrophication due to agricultural run off are noticed. Increased dependence on intensive agriculture and irrigation also results in salination, alkalization and water logging in irrigated areas of the country.

Deforestation: Forest is a renewable resource and contributes substantially to the economic development by providing goods and services to forest dwellers, people at large and forest based industries, besides generating substantial volume of employment. Forests are playing main role in enhancing the quality of environment by influencing the ecological balance and life support system.

Biodiversity: Biodiversity is one of the major environmental concerns in India, now India is in the tenth position in the world and fourth in Asia in plant diversity. The biodiversity in forests, non-irrigated lands, irrigated lands and hill areas deserts and marines ecosystems is subject to many pressures. One of the major causes of the loss of biological diversity has been the depletion of vegetative cover in order to expand agriculture. Since most of the biodiversity rich forests also contain the maximum mineral wealth and also the best sites for water impoundment, mining and development projects in such areas have led to destruction of habitats. Poaching and illegal trade of wildlife products too, have adversely affected biodiversity.

Pollution: The main factors contributing to urban air quality deterioration are growing industrialization and increasing vehicular pollution. It has been aggravated by developments that typically occur as countries industrialize, growing cities, increasing traffic, rapid economic development and industrial growth, all of which are closely associated with higher energy consumption. Industrial pollution is concentrated in industries like petroleum refineries, textiles, pulp and paper, industrial chemicals, iron and steel and non-metallic mineral products. Small scale industries especially foundries, chemical manufacturing and brick making are also significant polluters. In the power sector, thermal power, which constitutes bulk of the installed capacity for electricity generation, is an important source of air pollution.

Vehicle traffic is the most important source of pollution in all the mega cities. The number of vehicles in these cities has increased manifold. This increase has been characterized by a boom in private transport. Other reasons for high vehicular pollution are two stroke engines, aged vehicles, congested traffic, poor roads and outdated automotive technologies and traffic management system.

Coastal and marine pollutions are most contributing factors to environment. The coastal areas of India, with a coast line of over 7500 km. harbour a variety of specialized marine ecosystems like mangroves, coral reefs, salt lakes and mudflats which mainly form the habitat for endangered marine species and commercially important marine flora and fauna.

An important impact of climate change and global warming may be the rise in sea level. The primary effect of sea level rise will be increased coastal flooding, erosion, storm surges and wave activity. Primary studies on the impact of one-meter sea level rise on the Indian coastline indicate that 0.41 percent of India's coastal area will be inducted.

Poverty is also one of the reasons for environmental degradation in India. The circular link between poverty and environment is an extremely complex phenomenon. Inequality may foster unsustainability because the poor, who rely on natural resources more than the rich, deplete natural resources faster as they have no real prospects of gaining access to other type of resources. Acceleration in poverty alleviation is imperative to break this link between poverty and environment.

Impact of Environmental Degradation on Society

Environmental degradation is the result of the dynamic interplay of socioeconomic, institutional and technological activities. Environmental changes may be driven by many factors including economic growth, population growth, urbanization, intensification of agriculture, rising energy use and transportation. Poverty still remains a problem at the root of several environmental problems. The impact of environmental degradation on society has been discussed in the following.

It has been estimated that the process of deforestation, bad soil and water management, submergence of and in dam reservoirs, industrial and urban expansion, overgrazing, wind and water erosion, salination, flooding, water logging and so on, contribute to a loss of productivity in roughly one million hectare of land annually. The above process results in desertification and creation of drought prone conditions, leading to the immersion of those dependent on land for their daily subsistence.

Given these processes and the resultant, decline in livelihood for the millions critically dependent on these resources, there are few options but to cultivate increasingly marginal lands, thereby compounding the sustainability crisis. The impact on women is even more severe, as the loss access to fuel, fodder and water forces them to walk miles to collect the essential necessities for their subsistence.

The consequent escalation in the pressure on available arable land is so enormous that it has contributed to the growth of criminalization in the country side with illegal occupation of community lands, the formation of land armies by land lords to oppose any demands by marginal farmers for land, as well as increasingly militant movements to assert local control over productive resources like land and forests. The latter process most often results in state repression, compounding the climate of social unrest. Much of these have severally strained social relations within communities and between communities and the state. For instance, every year, over five lakh people are forcibly displaced by development projects alone. Most of them are not rehabilitated, and alternatives are rarely provided. In the process, communities and families are broken up, destroying structure of social and economic support.

Loss of cultural diversity is an inevitable consequence. For instance, a report based on a comprehensive survey of people displaced by the Rihan Dam stated that, "Many of the oustees, particularly the tribals, have fallen into the typical cycle of debt bondage, coupled with increasing destitution and intermittent employment as contract labourers in coal mines and elsewhere...most were simply kicked out with nothing left to fend for themselves".

Dimensions of Environmental Crisis in India

1. Of the 266 million hectares considered productive, about 175 million hectares are degraded in varying degrees (arid, alkaline, saline, waterlogged, ravine and the like). About 90 million hectare are acutely degraded, chiefly on account of loss of tree cover and top soil, leading to floods and drought.
2. Depletion of forest covers to about 19 percent of the total geographical area, instead of the desired 33 percent. India has only 2 percent of the forestland of the world but supports 16 percent of the world population.
3. Shortage of fuel wood and fodder for rural needs, leading to more pressure on the forests.
4. Threats to faunal and floral species and biological diversity, because of disturbance of their habitat.
5. Adverse impact of development activities such as mining, power generation, irrigation and industrialization.
6. Degradation of fragile ecosystems such as mangroves, irrigation lands, beaches and hill areas due to over exploitation, use of commercial agricultural practices, tourism and indiscriminate building activities.
7. Pollution of water from domestic and industrial waste.
8. Pollution of coastal areas and seas.
9. Air pollution due to emission from industries.
10. Increased production, transportation and use of hazardous chemicals.
11. Degradation of the urban environment because of rapid expansion and inadequate basic services.

Policies for Sustainable Development

Environmental problems like air pollution, water pollution,

soil degradation, deforestation, loss of bio-diversity, etc are caused by such diverse factors population growth, poverty, industrialization, agricultural development, transport development, urbanization, market failure etc. Such environmental degradation harms human health, reduces economic productivity and leads to the loss of amenities. Therefore, the damaging effects of environmental degradation can be reduced by a judicious choice of economic and environmental policies and environmental investments. The important policy measures for sustainable development are as follows.

Reducing Poverty: Reduction of poverty should be the foremost priority of the Government. It should select those projects which provide greater employment opportunities to the poor. It should expand health, family planning and education that will help reduce population growth. Supply of drinking water, sanitation facilities, and slum clearance should be given top priority.

Removing Subsidies: To reduce environmental degradation at no net financial cost to the Government, subsidies for resource use by the private and public sectors should be removed. Because, subsidies on the use of electricity, fertilizers, pesticides, diesel, petrol, gas, irrigation, water etc. lead to their wasteful use and environmental problems.

Clarifying and Extending Property Rights: Lack of property rights over excessive use of resources leads to degradation of environment. This leads to overgrazing, deforestation and over exploitation of minerals. Therefore, clarifying and assigning ownership titles to private owners will solve environmental problems

Market-based Approaches: Various market based approaches should be adopted to protect environment. Market based instruments in the form of emission tax, pollution taxes, marketable permits, depositor fund system, input taxes, differential tax rates, user administrative charges, subsidies for pollution abatement equipment etc. should be extensively used to protect environment.

Regulatory Policies: Regulatory policies are the other weapons for reducing environmental degradation. Regulators have to make decisions regarding price, quantity and technology. They decide the technical standards, regulations and charges on air, water and land pollutants.

Public Participation: Public awareness and participation are highly effective to improve environmental conditions. For this purpose, various formal and informal education programmes, environmental awareness programmes, advertisements, public movements, aforestation, conservation of wild life etc. should be organized on a large scale.

Trade and Environment: The Government should formulate an environment friendly trade policy covering both domestic and international trade. It should encourage the establishment of less polluting industries, adoption of cleaner technologies, adoption of environment friendly processes etc to control environmental degradation.

Participation in Global Environmental Efforts: Participation in various international conventions and agreements on environmental protection and conservation can also help to minimize damages of environmental degradation. They include the Montreal Protocol, the Basel Convention, the Rio Declaration, the Agenda 21, the Earth Summits, etc.

Renewable Energy: Policies should be framed for the use of renewable energy like solar and wind in place of coal and petrol. Atomic Energy Agency predicted that renewable energy would overtake natural gas to become the second largest source of power generation world wide with in two years, and that global wind and solar generating capacity would increase by more than 30 percent.

Global Environmental Issues

As early as 1896, the Swedish scientist Svante Arrhenius had predicted that human activities would interfere with the way, the sun interacts with the earth, resulting in global warming and climate change. His prediction has become true and climate change is now disrupting global environmental stability. The last few decades have seen many treaties, conventions, and protocols for the cause of global environmental protection.

Few examples of environmental issues of global significance are:

1. Ozone layer depletion;
2. Global warming; and
3. Loss of biodiversity.

Ozone Layer Depletion: Earth's atmosphere is divided into three regions, namely troposphere, stratosphere and mesosphere. The stratosphere extends from 10 to 50 kms. from the Earth's surface. This region is concentrated with slightly pungent smelling, light bluish ozone gas. The ozone gas is made up of molecules each containing three atoms of oxygen; its chemical formula is O3. The ozone layer, in the stratosphere acts as an efficient filter for harmful solar Ultraviolet B (UV-B) rays. Ozone is produced and destroyed naturally in the atmosphere and until recently, this resulted in a well-balanced equilibrium. Ozone is formed when oxygen molecules absorb ultraviolet radiation with wavelengths less than 240 nanometres and is destroyed when it absorbs ultraviolet radiation with wavelengths greater than 290 nanometres. In recent years, scientists have measured a seasonal thinning of the ozone layer primarily at the South Pole. This phenomenon is being called the ozone hole.

- **Effects of Ozone Layer Depletion on Human and Animal Health:** Increased penetration of solar UV-B radiation is likely to have high impact on human health with potential risks of eye diseases, skin cancer and infectious diseases.
- **Effects on Terrestrial Plants:** In forests and grasslands, increased radiation is likely to change species composition thus altering the bio-diversity in different ecosystems. It

could also affect the plant community indirectly resulting in changes in plant form, secondary metabolism, etc.

- **Effects on Aquatic Ecosystems:** High levels of radiation exposure in tropics and subtropics may affect the distribution of phytoplanktons, which form the foundation of aquatic food webs. It can also cause damage to early development stages of fish, shrimp, crab, amphibians and other animals, the most severe effects being decreased reproductive capacity and impaired larval development.
- **Effects on Bio-geo-chemical Cycles:** Increased solar UV radiation could affect terrestrial and aquatic bio-geo-chemical cycles thus altering both sources and sinks of greenhouse and important trace gases, e.g. carbon dioxide (CO_2), carbon monoxide (CO), carbonyl sulfide (COS), etc. These changes would contribute to biosphere-atmosphere feedbacks responsible for the atmosphere build-up of these greenhouse gases.
- **Effects on Air Quality:** Reduction of stratospheric ozone and increased penetration of UV-B radiation result in higher photo dissociation rates of key trace gases that control the chemical reactivity of the troposphere. This can increase both production and destruction of ozone and related oxidants such as hydrogen peroxide, which are known to have adverse effects on human health, terrestrial plants and outdoor materials.
- The ozone layer, therefore, is highly beneficial to plant and animal life on earth filtering out the dangerous part of sun's radiation and allowing only the beneficial part to reach earth. Any disturbance or depletion of this layer would result in an increase of harmful radiation reaching the earth's surface leading to dangerous consequences.

Ozone Depletion Counter Measures:

- International cooperation, agreement (Montreal Protocol) to phase out ozone depleting chemicals since 1974.
- Tax imposed for ozone depleting substances.
- Ozone friendly substitutes-HCFC (less ozone depleting

potential and shorter life).

- Recycle of CFCs and Halons.

Global Warming: Before the Industrial Revolution, human activities released very few gases into the atmosphere and all climate changes happened naturally. After the Industrial Revolution, through fossil fuel combustion, changing agricultural practices and deforestation, the natural composition of gases in the atmosphere is getting affected and climate and environment began to alter significantly. Over the last 100 years, it was found out that the earth is getting warmer and warmer, unlike previous 8000 years when temperatures have been relatively constant. The present temperature is 0.3-0.6°C warmer than it was 100 years ago.

The key greenhouse gases (GHG) causing global warming is carbon dioxide. CFCs, even though they exist in very small quantities, are significant contributors to global warming. Carbon dioxide, one of the most prevalent greenhouse gases in the atmosphere, has two major anthropogenic (human-caused) sources: the combustion of fossil fuels and changes in land use. Net releases of carbon dioxide from these two sources are believed to be contributing to the rapid rise in atmospheric concentrations since Industrial Revolution. Because estimates indicate that approximately 80 percent of all anthropogenic carbon dioxide emissions currently come from fossil fuel combustion, world energy use has emerged at the centre of the climate change debate.

Global Warming (Climate Change) Implications:

- **Rise in Global Temperature:** Observations show that global temperatures have risen by about 0.6°C over the 20th century. There is strong evidence now that most of the observed warming over the last 50 years is caused by human activities. Climate models predict that the global temperature will rise by about 6°C by the year 2100.
- **Rise in Sea Level:** In general, the faster the climate change, the greater will be the risk of damage. The mean sea level is expected to rise by 9-88 cm. by the year 2100,

causing flooding of low lying areas and other damages.

- **Food Shortages and Hunger:** Water resources will be affected as precipitation and evaporation patterns change around the world. This will affect agricultural output. Food security is likely to be threatened and some regions are likely to experience food shortages and hunger.
- **India could be more at risks than many other countries:** Models predict an average increase in temperature in India of 2.3 to 4.8° C for the benchmark doubling of carbon-dioxide scenario. Temperature would rise more in Northern India than in Southern India. It is estimated that 7 million people would be displaced, 5,700 km. of land and 4200 km. of road would be lost, and wheat yields could decrease significantly.

Loss of Biodiversity: Biodiversity refers to the variety of life on earth, and its biological diversity. The number of species of plants, animals, micro organisms, the enormous diversity of genes in these species, the different ecosystems on the planet, such as deserts, rainforests and coral reefs are all a part of a biologically diverse earth. Biodiversity actually boosts ecosystem productivity where each species, no matter how small, all have an important role to play and that it is in this combination that enables the ecosystem to possess the ability to prevent and recover from a variety of disasters.

It is now believed that human activity is changing biodiversity and causing massive extinctions. The World Resource Institute reports that there is a link between biodiversity and climate change. Rapid global warming can affect ecosystems chances to adapt naturally. Over the past 150 years, deforestation has contributed an estimated 30 percent of the atmospheric build-up of CO_2. It is also a significant driving force behind the loss of genes, species, and critical ecosystem services.

Link between Biodiversity and Climate Change

1. Climate change is affecting species already threatened by

multiple threats across the globe. Habitat fragmentation due to colonization, logging, agriculture and mining etc. are all contributing to further destruction of terrestrial habitats.

2. Individual species may not be able to adapt. Species most threatened by climate change have small ranges, low population densities, restricted habitat requirements and patchy distribution.
3. Ecosystems will generally shift northward or upward in altitude, but in some cases they will run out of space–as 10°C change in temperature correspond to a 100 km. change in latitude, hence, average shift in habitat conditions by the year 2100 will be on the order of 140 to 580 kms.
4. Coral reef mortality may increase and erosion may be accelerated. Increase levels of carbon dioxide adversely impact the coral building process (calcification).
5. Sea level may rise, engulfing low-lying areas causing disappearance of many islands, and extinctions of endemic island species.
6. Invasive species may be aided by climate change. Exotic species can out-compete native wildlife for space, food, water and other resources, and may also prey on native wildlife.
7. Droughts and wildfires may increase. An increased risk of wildfires due to warming and drying out of vegetation is likely. Sustained climate change may change the competitive balance among species and might lead to forests destruction.

Kyoto Protocol

There is a scientific consensus that human activities are causing global warming that could result in significant impacts such as sea level rise, changes in weather patterns and adverse health effects. As it became apparent that major nations such as the United States and Japan would not meet the voluntary stabilization target by 2000, Parties to the Convention decided in 1995 to enter into negotiations on a protocol to establish legally binding limitations or reductions in greenhouse gas emissions. It was decided by the Parties that this round of negotiations would establish limitations only for the developed countries, including the former Communist countries (called annex A countries).

Negotiations on the Kyoto Protocol to the United Nations Framework Convention on Climate Change (UNFCCC) were completed December 11, 1997, committing the industrialized nations to specify, legally binding reductions in emissions of six greenhouse gases. The 6 major greenhouse gases covered by the protocol are carbon dioxide (CO_2), methane (CH_4), nitrous oxide (N_2O), hydro fluorocarbons (HFCs), per fluorocarbons (PFCs), and sulfur xafluoride (SF_6).

Mr. Ban Ki-moon, UN Secretary General, who received the Sustainable Development Leadership Award 2009 at the Delhi Sustainable Summit said "a balanced, comprehensive and effective" international agreement will take place by the end of December 2009 at Copenhagen. As a regime, it must also be "ratifiable" as a successor to the Kyoto Protocol.

Mr. Ban said eradication of poverty is not possible if we neglect or deplete our natural capital. He said science had shown that we were depleting the planet's natural assets at an unsustainable rate. Failure to combat climate change would increase poverty and hardships, destabilise economies, and breed insecurity in many countries.

Conclusion

Rapid environmental degradation that has been taking place all over the world in recent decades has alarmed economists and environmentalists alike. Fostering sustainable development requires the effective management of naturally human and physical capital. Improved coordination across the countries to share the global resources, technology and also scarce resources has become the need of the hour. Global level generosity in promoting and protecting democracy, exchange of technology, maintaining stability of prices in various economies, judicious use of all environmental material throughout to enhance human development and sustainable development can be achieved only if the environment is conserved and improved.

References

Chopra Kanchan and Kadekodi G.K. (1999), 'Operationalising Sustainable Development: Economic-Ecological Modelling for Developing Countries' Concept Publishers, New Delhi.

Jayanta Bandyopadhyay and Bidisha Malik (2003), 'Ecology and Economics in Sustainable Water Resources Development in India', in Kanchan Chopra et al. (eds.) 'Water Resources, Sustainable Livelihoods and Eco-system Services', Concept Publishers, New Delhi.

Krishnamurthy, H.R. (2005), 'Economic Development of India', Sapna Publishers, Bangalore.

Misra, S.K. and Puri, V.K. (2005), 'Indian Economy', Himalaya Publishing House, New Delhi.

Prakash, H. et al. (2005), 'Environment and Sustainable Development', Southern Economist, 43(21).

Sengupta Ramprasad (2001), 'Ecology and Economics: An Approach to Sustainable Development', Oxford University Press, India.

Viegas, Philip, Menon, Geetha (1989), 'The Impact of Environmental Degradation on People', Indian Social Institute, New Delhi.

5

Land Acquisition for Special Economic Zones (SEZs)

M. Yadagiri and N. Sreenivas

Introduction

Special Economic Zone (SEZ) is a geographical area that has economic laws which are more liberal in a country than the normal economic laws. The first of its kind was introduced in the 80s by the Government of China. The most successful SEZ in China was Shenzhen Village which turned out to be a city wherein the population rose to 10 million within 20 years. The concept of SEZ was followed by other developing countries from 1990s onwards including countries like Russia, Philippines, North Korea, Ukraine, India, Iran, Poland, Kazakhstan and Jordan, etc. In United States, SEZ is called as Urban Enterprise Zone (UEZ). It is being estimated that there are about 3000 SEZs commenced till the year 2007 spreading over approximately 120 countries. The Special Economic Zone Act, 2005 received the assent of the President of India on 23rd June 2005 to provide for the establishment, development and management of the Special Economic Zones.

Special Economic Zones are treated as a foreign territory for various purposes such as tariffs, trade, and duties. For instance, an industrialist will have to pay no import duties or excise for goods imported and produced there. However, they are required to export only one percent of products produced in the SEZ. In addition, they will get cent percent income tax exemption for 5 years. For the next 5 years, they are required to pay only half the tax if profits are ploughed back in the SEZ. Apparently, the Government of India has made several policies with regard to SEZs in the country and encourages the setting up of SEZs in the country, for they help in the economic and industrial growth of a nation. According to the Government of India's policy, a SEZ in the country has to be built on at least 1 thousand hectares or more of land. And so Land Acquisition on a massive scale is taking place in India so that more and more SEZs can be set-up in the country.

Special Economic Zones Act, 2005

India was one of the first in Asia to recognize the effectiveness of the Export Processing Zone (EPZ) model in promoting exports, with Asia's first EPZ set up in Kandla (Gujarat) in 1965. However, the EPZs were not able to emerge as effective instruments for export promotion on account of multiplicity of controls and clearances, absence of world-class infrastructure and an unstable fiscal regime.

Considering the need to enhance foreign investment and promote exports from the country and realising the need that level playing field must be made available to the domestic enterprises and manufacturers to be competitive globally, the Special Economic Zones (SEZs) Policy was announced in April 2000. This policy intended to make SEZs an engine for economic growth supported by quality infrastructure complemented by an attractive fiscal package, both at the Centre and the State level, with the minimum possible regulations. SEZs in India functioned from November 1, 2000 to February 9, 2006 under the provisions of the Foreign Trade Policy and fiscal incentives were made effective through the provisions of relevant statutes.

To instil confidence in investors and signal the Government's commitment to a stable SEZ policy regime and with a view to impart stability to the SEZ regime thereby generating greater economic activity and employment through the establishment of SEZs, a comprehensive draft SEZ Bill was prepared after extensive discussions with the stakeholders. The Special Economic Zones Act, 2005, was passed by Parliament in May 2005 which received Presidential assent on

June 23, 2005. After extensive consultations, the SEZ Act, 2005, supported by SEZ Rules, came into effect on February 10, 2006, providing for drastic simplification of procedures and for single window clearance on matters relating to Central as well as State Governments. The main objectives of the SEZ Act are the following:

1. Generation of additional economic activity.
2. Promotion of exports of goods and services.
3. Promotion of investment from domestic and foreign sources.
4. Creation of employment opportunities.
5. Development of infrastructure facilities.

The SEZ Act, 2005 envisages key role for the State Governments in export promotion and creation of related infrastructure. A single window SEZ approval mechanism has been provided through a 19 member inter-ministerial SEZ Board of Approval (BoA). The applications duly recommended by the respective State Governments/UT Administrations are considered by BoA periodically. All decisions of the BoA are with consensus.

The SEZ Rules provide for differentiated minimum land requirements for different classes of SEZs. Every SEZ is divided into a processing area where alone the SEZ units would come up and the non-processing area where supporting infrastructure is to be created. The SEZ Rules also provide for simplified procedures for development, operation and maintenance of the SEZ, setting up units in SEZs, single window clearance both relating to Central as well as State Governments for setting up of an SEZ and units in a SEZ, and simplified compliance procedures/documentation with emphasis on self-certification. A Board of Approval has been constituted by Government in exercise of the powers conferred under the SEZ Act.

SEZs in India seek to promote value addition component in exports, generate employment and mobilize foreign exchange. SEZs when operational are expected to offer high quality infrastructure facilities and support services, besides allowing for the duty free import of capital goods and raw materials. Additionally, attractive fiscal incentives and simpler customs, banking and other procedures are offered in such zones. Setting up of SEZs is also treated as an infrastructure development activity and offered same incentives. SEZs in India closely follow the Chinese model.

Approval Mechanism and Administrative Set up of SEZs

Approval Mechanism: The developer submits the proposal for establishment of SEZ to the concerned State Government. The State Government has to forward the proposal with its recommendation within 45 days from the date of receipt of such proposal to the BoA. The applicant also has the option to submit the proposal directly to the BoA.

BoA has been constituted by the Central Government in exercise of the powers conferred under the SEZ Act. All the decisions are taken in the BoA by consensus. BoA has 19 Members. It is headed by the Secretary, Department of Commerce in the Ministry of Commerce and Industry, Government of India.

Administrative Set up: The functioning of the SEZs is governed by a three- tier administrative set up. BoA is the apex body and is headed by the Secretary, Department of Commerce. The Approval Committee at the Zone level deals with approval of units in the SEZs and other related issues. Each Zone is headed by a Development Commissioner, who is ex-officio chairperson of the Approval Committee.

Once an SEZ has been approved by the BoA and Central Government has notified the area of the SEZ, units are allowed to be set up in the SEZ. All the proposals for setting up of units in the SEZ are approved at the Zone level by the Approval Committee consisting of Development Commissioner, Customs Authorities and representatives of State Government. All post-approval clearances including grant of importer-exporter code number, change in the name of the company or implementing agency, broad banding diversification etc. are given at the Zone level by the Development Commissioner. The performance of SEZ units is periodically monitored by the Approval Committee and units are liable for penal action under the provision of Foreign Trade (Development and Regulation) Act, 1992 in case of violation of the conditions of the approval.

Criteria for Approval

Proposals for setting up SEZ in the public/private/joint/State sector are required to meet the following conditions.

1. Minimum size of the SEZ shall not be less than 1000 hectares. This would however, not apply to existing EPZs converting into SEZs as such or for notifying additional area as a part of such SEZ or to product-specific port/airport based SEZs.
2. SEZ and units therein shall abide by local laws, rules, regulations or bye-laws in regard to area planning, sewerage disposal, pollution control and the like. They shall also comply with industrial and labour laws and such other laws/rules and regulations as may be locally applicable.
3. SEZs shall make adequate arrangements to fulfil all the requirements of the laws, rules and procedures applicable to such SEZs.
4. Only units approved under the SEZ scheme would be permitted to be located in SEZ.

Proposals received for setting up of SEZs are considered by an Inter-Ministerial Committee known as the Board of Approval (BoA), in its meeting held every month. The promoters of the SEZs are also invited to such meetings to explain the salient features of their proposals and to furnish clarifications, wherever required.

Proposals that are received up to 15 days prior to the scheduled meeting of the Board of Approval are placed for its consideration and any proposals received thereafter are considered in the next meeting.

Conversion of EPZs into SEZs

How are SEZs Different from EPZs? In one sense, SEZs are only a variant of EPZs. Both have delineated area and enjoy duty free import of capital goods and raw materials. Both aim to attract foreign investment for setting up export-oriented units by providing developed infrastructure, conducive operating environment and a package of fiscal incentives.

However, the objectives of SEZs are much larger than mere promotion of export processing activities. While EPZs are industrial estates, SEZs are virtually industrial townships that provide supportive infrastructure such as housing, roads, ports and telecommunications. The scope of activities that can be undertaken in the SEZs is much wider and their linkages with the domestic economy are stronger. Consequently, they have a diversified industrial base. Their role is not transient like that of EPZs, as they are intended to be instruments of regional development as well as export promotion. As such, SEZs can have tremendous impact on exports, inflow of foreign investment and employment generation.

Indian EPZs have not fared well as these are so few and their areas are small. For some, the location is inappropriate and they are subject to cumbersome procedures for long. However, in many East Asian economies, EPZs have been highly successful in their initial phase of opening up. Therefore, the mere fact that India is switching over to the SEZ model may not per se guarantee success as demonstrated by China unless policy parameters and implementation strategy are planned carefully.

Some of the existing EPZs have been converted into SEZs. For example, the Government has converted EPZs located at Kandla and Surat (Gujarat), Cochin (Kerala), Santa Cruz (Mumbai-Maharashtra), Falta (West Bengal), Chennai (Tamil

Nadu), Visakhapatnam (Andhra Pradesh) and NOIDA (Uttar Pradesh) into SEZs.

Incentives and Facilities Offered to Units in SEZs

The incentives and facilities offered to the units in SEZs for attracting investments into the SEZs, including foreign investment, include the following:

1. Duty free import/domestic procurement of goods for development, operation and maintenance of SEZ units.
2. 100 percent income tax exemption on export income for SEZ units under Section 10AA of the Income Tax Act for first 5 years, 50 percent for next 5 years thereafter and 50 percent of the ploughed back export profit for next 5 years.
3. Exemption from minimum alternate tax under Section 115JB of the Income Tax Act.
4. External commercial borrowing by SEZ units up to US$ 500 million in a year without any maturity restriction through recognized banking channels.
5. Exemption from Central Sales Tax.
6. Exemption from service tax.
7. Single window clearance for Central and State level approvals.
8. Exemption from State sales tax and other levies as extended by the respective State Governments.

Incentives and Facilities Available to SEZ Developers

These include the following:

1. Exemption from customs/excise duties for development of SEZs for authorized operations approved by the BoA.
2. Income tax exemption on income derived from the business of development of the SEZ in a block of 10 years in 15 years under Section 80-IAB of the Income Tax Act.
3. Exemption from minimum alternate tax under Section 115 JB of the Income Tax Act.
4. Exemption from dividend distribution tax under Section 115O of the Income Tax Act.

5. Exemption from Central Sales Tax (CST).
6. Exemption from Service Tax (Sections 7 and 26 and Second Schedule of the SEZ Act).

Land Requirements to Set Up SEZs

The total land requirement for the formal approvals granted till date is approximately 67,772 hectares out which about 109 approvals are for State Industrial Development Corporation/State Government Ventures which account for over 20,893 hectares. In these cases, the land already available with the SIDBs/State Governments or with Private Companies has been utilized for setting up SEZs. The land for the 260 notified SEZs where operations have since commenced involved approximately over 29,953 hectares only. Out of the total land area of 29,73,190 sq. km. in India, total agricultural land is of the order of 15,34,166 sq. km.. It is interesting to note that out of this total land area the land in possession of 260 SEZs notified amounts to approximately 299 sq. km. only. The formal approvals granted also works out to only around 677 sq. km.

Purpose of Land Acquisitions

Land acquisition literally means acquiring of land for some public purpose by Government or its agency, as authorized by the law, from the individual land owner's after paying Government fixed compensation in lieu of losses incurred by land owners due to surrendering of their land to the concerned government agency. The purpose of land acquisition may be for putting up educational institutions, schemes such as housing, health or slum clearance etc. for the development of rural area and for the development of agricultural or industries.

Land Acquisition Proceedings for SEZs

Indian expropriation law acquired corporate dimension in the 1960s, when under Section 55 of the Land Acquisition Act,

Centre issued rules for initiating acquisition proceedings on behalf of companies. The applicant had to fulfil certain conditions before the process was initiated, as having made best efforts to identify suitable land, and failed in negotiating a reasonable price. The land had to be suitable for the purpose of acquisition and agricultural land acquisition would be considered, only if there was no alternative. The Collector had to determine this in consultation with the district's senior agricultural officer, with public purpose as the ultimate objective of the acquisition. The company was bound by strict compliances and obligations, a breach could entail reversion of the land to the government. After the initial notification and hearing of objections, the Collector was required to prepare a report, and in case of a company, provide reasons as to why the land had to be acquired, and if it was good agricultural land, deal with the objections in accordance with the above rules. If the government was not funding the acquisition, the Collector did not have to justify it against the touchstone of public purpose.

Issues and Challenges of Land Acquisition in India

SEZs and Land Acquisition are inter-connected and SEZs have picked up speed in India since the Indian government has encouraged the setting up of SEZs in the country. Acquiring the land is the biggest incentive held out to private SEZ developers by the state governments. The state involvement becomes more apparent in the larger zones that are in the land acquisition phase. Governments, increasingly challenged over their roles in acquiring land for private development, argue that SEZs are needed for the 'development' of their states, and that they need to do everything possible to attract the promoters of these zones to their own state. Whether SEZs will bring in the promised benefits is an entirely different discussion. What is examined here is the injustice of using the colonial land acquisition law that has become such a favourite instrument in the hands of state governments.

Large-scale land acquisition for development will involve acquiring settled lands with long-established rights. It will disrupt the life of some functioning community, and of all who live in it, not just the land-owners. In fact, the ones worst affected will be the share-croppers and labourers, the petty traders and service providers. These landless ones do not even have a juridical basis for compensation, if the transaction is seen simply as a sale of land, voluntary or compulsory.

Land acquisition for SEZs is taking place mainly in agricultural lands and the central and state governments are acquiring the land from the farmers. Across India, the total amount of land, which will be acquired, is around 150,000 hectares which is capable of producing around 1 million tons of agricultural produce.

The distinction between land acquisition for a company and for public purpose was made in the 1984 amendments, which specifically excluded land acquisition for firms from the definition of public purpose. Section 6, of the Act, which requires a declaration by the government that a particular land is needed for public purpose, clarifies that no such declaration is required in case of companies, unless any part of the compensation is paid out of public funds. The conclusion drawn from this is that while other acquisitions are for public purpose, in case of a company, provided it meets the entire cost of acquisition, no such justification is required. Going by the 1963 rules, a company can seek acquisition, if it has failed to renegotiate a reasonable price otherwise. But would that absolve companies from commitment to public purpose under the rules. Public purpose is involved even in acquisition for establishing an industry by the private sector, and no government can be faulted for facilitating acquisition for investors.

The SEZ Act is criticized as an imperial legislation, lacking ameliorative measures for the deprived, with obsolete benchmarks for determining compensation. The Act envisages compensation parameters to include market value of property on date of preliminary notification, damages suffered by "interested" person(s) on account of his earnings, other properties, etc., reasonable expenses for relocation, 12 percent interest up to date of handing over of possession, and a 30 percent solatium on the above. But "market price" is neither properly defined nor regulated, and therefore, very often not justified. There is a serious lack of coordination between the various arms of the government in this exercise-rehabilitation, commerce and industry, and rural development. And it's important not to lose sight of the 234 SEZs cleared without acquisition issues, the investments and jobs involved.

The total number of SEZ to come up in the near future in India is about 234 wherein there will be a total investment of ₹ 3,00,000 crore till December 2009. A total of 87 SEZs have been sanctioned in the country which has attracted investment to the tune of ₹ 60,000 crore. If all the 234 SEZs were allowed to be cleared, it may employ about 40,00,000 people in India by the end of December 2009. However, on the other hand, the proposed SEZs may require maximum of about 1750 sq. km. land when the total land in India constitutes 29,73,190 sq. km., out of which 15,34,166 sq. km. is being used as the agricultural land.

The central issue that has to be addressed is not the encroachment on land for agriculture, but displacement of people. The numbers displaced by development projects in post-independence India are not known because the government has not released any reliable figures on this. Informally, estimates put the number at 40-50 million. Almost all of these people are involuntary displacers. That is because we have relied for too long on the Land Acquisition Act of 1894, which allows the government to exercise the right of eminent domain for compulsorily acquiring private lands for public purposes.

Numerous issues arise across the country from Barnala in Punjab and Jhajjar in Haryana to Kakinada in Andhra Pradesh and Nandagudi in Karnataka, from Nandigram in West Bengal and Jagat Singh Pur in Orissa, in the East from Raigad, Mangalore and Goa to the West coast of India, farmers, landless-workers, fish-workers and artisans have expressed their anger against the loss of land, livelihood and habitat. In some cases, as in Nandigram and in Goa, projects have been cancelled. In some cases, SEZ land is being de-notified, despite the SEZ Act. In Goa, people's struggle has been so successful that the SEZ policy has been abandoned by the government recently. The West Bengal Government acquired level fertile agricultural land in West Medinipur for Tata Metaliks in 1992, leading to dispossessing small and marginal farmers, in preference to undulating wasteland that was available nearby.

In the case of the Century Textiles Pig Iron Plant in the same area, the state government acquired about 525 acres of land for a proposed plant in 1996. However, till 2003 the factory had not come up and neither had all the original land owners been fully compensated. The company had decided that pig iron production was no longer profitable and refused to pay the compensation and take over the land. Singur is another example of the same government acquiring prime agricultural land for a car factory, The Tata Motors. State governments have not hesitated to acquire land even (mis)using draconian emergency powers available in the Land Acquisition Act.

A case in point is the Tamil Nadu government's acquisition of land near Pulicat Lake north of Chennai for a petro-chemical complex. Recently, due to Land Acquisition in Atchutapuram, near Visakhapatnam, the farmers have declared war against the setting up of SEZ in their lands. The land compensation money that the state government is giving is very less in comparison to the actual market rate of the land, and this has angered the farmers even more. This is bound to inspire workers and peasants in other states to continue and accelerate their protests against SEZs to protect their livelihood. Thus, SEZs have become a burning issue and are currently under the scanner.

It is estimated that more than 10 lakh people who are dependent upon agricultural lands will be evicted from their lands and the farming families will have to face loss of around ₹ 212 crore each year in total income. It will also lead to putting the food security of India at risk. SEZs and Land Acquisition in India has now resulted in dissent, uproar, and opposition from the farmers, for their livelihood has been put at stake.

Suggestions

SEZs and Land Acquisition has been taking place in India at a very fast pace over the last few years. The Government of India must make sure that Land Acquisition and SEZs must prove beneficial for the people of the country and not harmful. The following are the suggestions to overcome from these challenges to succeed in establishing SEZs in India:

1. The government should oppose to forcible acquisition of land for setting up Special Economic Zone (SEZ) in the country. Lands belonging to the farmers must not be sold to other parties. Instead, the ownership of the land must remain with the farmers themselves while the government could take it on lease from the farmers.
2. The policy framework of Special Economic Zones (SEZs) is getting tweaked to meet the aspirations of all stake-holders. As per the new norms, the size of a SEZ cannot exceed 12,500 acres, when earlier there was a lower limit of multi-product SEZs which can be formed on 2,500 acres of land with no limit to the upper. It has also been decided that the state government can no longer acquire land for the SEZs for the private developers.
3. For industrialization, land is needed; however, those who are losing their land must be compensated properly.
4. It is essential to look at the costs and benefits of setting up an industry from the point of view of overall economic policy.

5. For large-scale industries, private and public, very strict environmental and land-use planning regimes are needed so that fertile land cannot be destroyed or hazardous industries cannot pollute the environment, atmosphere, land and water system.
6. On the issue of land acquisition for public purposes, there have been inadequacies in compensation, and in ensuring that interest of all stake-holders who suffer in this process should be considered and the reconciliation of land rights and development is required.

6

Environmental Stress and Gender

K. Hari Babu and K. V. R. Srinivas

In developing nations, issues of survival are inextricably linked with environment for a majority of population. This is especially true for women who interface extensively with the environment which directly affects their physical, social and psychological well being and more so their quality of life. India is going through a period of rapid changes. The economic, political and social changes will affect the lives of all individuals and especially women. In the shift from traditional society to modern operation, developmental processes in India have neglected both women and the environment. Women play a predominant role in activities related to natural resources and domestic sphere. There are drawers of water, hewers of wood, labourers. Preparers of food, bearers of children, educators, health care providers and decision makers. Although they are central to caring for families and communities, to production and reproduction, they have accorded unequal status. Throughout the world they are over burdened and undervalued. Their subordination makes it difficult for them to cope with various demands made upon them whether on a physical or social or emotions nature (WHO, 1993).

More so, there is a strong line of thought that emphasizes women's special ecological consciousness, which makes them better managers of natural resources. The close association between women and natural resources is then valid primarily in rural context, where because of socio-economic roles through generations; women are required to provide food, water, fuel, fodder and income from the surrounding resource-base for survival.

It is observed that women's perceptions and motivations are circumscribed by the domestic sphere at life where they have lots roles and responsibilities related to nurturing and home making. Fulfilling the roles and responsibilities of domestic sphere is dependent on the surrounding resource base from where they collect food, water, fuel, fodder and minor forest produce. If resource-base gets degraded, it subsequently interferes with women's responsibility towards domestic sphere, which negatively affects women, physically as well as psychologically (Parul Rishi, 1997).

Environmental Stress and Gender

Environmental stress is a process that occurs when there is an imbalance between environmental demands and response capabilities of individual (Lazarus, 1966). The environmental events which initiate this stress process are called stressors and the reaction caused by the stressors is called the stress response, which is characterized by emotional changes, behaviour directed toward reduction of stress and the physiological changes such as increased arousal. Rural population is especially exposed to number of stressors existing in the natural or social environment because of their great physical as well as psychological dependence on the surrounding environment.

Women are more vulnerable than men to the negative effects of various stressors which are often exacerbated by material environment, social environment, financial environment and more so, their own psychological frame of mind i.e. the cognitive appraisal. There is a growing recognition that the stresses imposed on women affect their physical, emotional and mental well being.

Physical Stressors

Stresses which are existing in the natural environment are sometimes the result of development and population growth and they adversely affect the people dependent on that environment in general and women in particular. Major examples of the physical stressors may include natural resource depletion, environmental degradation, pollution of existing resources, in door air pollution and atmospheric pollution caused by transport and industry. All these physical stressors have subsequent effects on women and adversely affect their health—because women in rural subsistence economies are the main providers of food, fuel and water and the primary caretakers of their families. They depend heavily on community owned croplands, grasslands and forests to fulfil their familial needs. The widespread depletion and degradation of these resources has led to equally widespread improvement of subsistence families throughout the developing world.

Taken specifically, resource depletion and environmental degradation have consequences on women time, income, nutrition, health and social support network. Now, they will have to go greater distances for collecting natural resources which takes away greater time and energy Besides, their income from the sale of non timber forest products is also reduced, causing greater financial stress. The major part of the diet of rural people used to be the gathered foods from forests like fruits, vegetables, nuts, seeds and oils which are being reduced causing nutritional deficiency specifically in women who require it much being the child bearers. Pollution of rivers and ponds with fertilizers and pesticides also affect the health of expectant mothers and their babies. When resources are depleted, males of the family migrate to other places for employment and again, Women are the sufferers because at one hand, they loose the social support and at another hand their work load further increases.

Another form of physical stress is indoor air pollution which is caused by biomass fuel used for cooking and heating in rural areas by women and young girls. Biomass fuel in the form of wood, dung and crop residues, limits large number of pollutants like suspended particles, carbon monoxide, nitrogen oxides, formaldenhyde, benjopyrene (bap) and compounds such as polyaromatic hydrocarbons. In a study, conducted in four villages of Anand district, Gujarat (CSE, 1985), average exposure of women to suspended particulate ranged from 1,110 to 56,000 micro programmes per cubic meter as compared to 120-150 micro programmes recommended by WHO. When dung is added to wood concentrates increase further.

Measurement of bap revealed exposure levels equivalent to smoking about 20 packs of cigarettes per day. Besides, poor Ventilation of kitchens increases the *exposure* to pollutants several times more (CSE, 1985).

A study conducted by Tata Energy Research Institute (TERI, 1992) in villages of Pauri district of Garhwal, found that women faced the highest daily exposure to both TSP and Co. In a report of WHO (1991), prolonged exposure to such high levels of pollutants has a very adverse impact on women's health. The effects may include eye problems, respiratory problems and adverse reproductive outcomes.

Gender Bias: A Social Psychological Stressor

Women perform the lion's share of work in subsistence economies toiling longer hours and contributing more to family income than their male relatives.but are viewed as "unproductive" in the eyes of statisticians, economists, development experts and even their husbands. A huge proportion of the world's real productivity therefore, remains undervalued, and the essential contributions women make to the welfare of families and nations remains unrecognized.

Gender bias is a world wide phenomenon but it is especially pernicious in the third world, where most of the women's activity takes place for the purpose of house hold consumption. Gender bias is one of the major psycho-social stressor affecting women. It ranges from exclusion of women from developmental programmes to wage discrimination and systemic violence against females. There is a gross unequal allocation of resources–whether of food education, training, jobs, information or rights. Women lack the ownership of land and there are gender specific restrictions on land use pattern of females. They lack the material resources and have the very limited or no rights over come or cash resources. Another stressor for women is the greater work load of fields as well as house making and on the other hand male control over decision making and participation in any community activity. A very limited attention and recognition is given to their viewpoint or needs and their viewpoint or needs and their traditional knowledge is rarely recognized in the male dominated society.

Besides, in the family also, they are the ultimate sufferers because financial stress in the family or the domestic violence or any other family problem which is ultimately the stressor for women, being the predominant care giver and homemaker in the family.

Stress and Women's Health

There is a growing recognition that the stressors imposed on women affect their physical, emotional and psychological well being. Epidemiological evidence links mental and physical disorders with alienation, powerlessness, over work strain, subordination and undervaluation, the conditions most frequently experienced by most of the women. Considerable research has revealed that women experience and respond to stress in distinctive ways compared to men. Women's stress response process is both qualitatively and quantitatively different in terms of hormonal profiles and in emotional quality. In addition, the nature of women's lives and realities renders them at risk for stress related efforts more often than men. This reflects the greater number of social roles women fill as wife, mother, daughter, employee and care of others.

Beyond that, women's reproductive role of bearer, producer, feeder and nurturer of children produced unique potential for stress related effects. Only perhaps so single fathers with dependent children approach sheer level of multiple responsibilities that the majority of women carry. It is clear that those who have particular personal difficulties along with gender specific stresses, succumb at some point or another (WHO/FHE/MNH, 1993).

Very often she is poor, malnourished, uneducated, overworked, socially deprived, having poor physical health with anaemia and repeated pregnancies. She also has mental distress resulting in anxiety, depression, hysteria, somatisation and many other mental disorders. There is perhaps no single group that illustrates better the combined impact of poverty, lack of education, unemployment and social disintegration on health and quality of life than women. Throughout the world, women are overworked and undervalued. Their subordination makes it more difficult for them to cope with demands made on them whether of a physical, social or emotional nature. Thus, serious attention is to be given to the stress management and coping aspects of women's life in particular, to safeguard them from the further damages to their physical and mental health.

Stress Management

"Indian women, while sharing many of the problems common to other sisters in other parts of the world, face certain unique ethnic and cultural problems. Today they are awake and alert but which direction she will go will not depend on just her own inclination and efforts but also as her family, society, the education and health services and sectors, the laws and above all what her enlightened brothers and sisters have to offer her at this crucial juncture as a practical package" (Sathyavathi, 1995).

In an effort to manage the everyday increasing environmental stressors among women in particular, an integrated package of stress management is required in which efforts should be initiated on the part of government, voluntary organizations, society, community, family as well as the women themselves.

Counteracting Gender Bias

Gender bias is a world wide phenomenon which imposes several discriminations on women like little access to productive resources, little control over family income, and dependence on husband and children for social status and economic security. All such discriminations lead to feelings of subordination, powerlessness, dependence, psychological stress and depression. To avoid all such negative outcomes of gender-bias, there is a need to inculcate self sufficiency and independence among women, Voluntary organizations working in the rural areas for the participation and empowerment of women in the development sector should *first of* all make *their* special efforts towards "Psychological empowerment" of women because females are socialized in a dependent and subordinated manner. It is observed that, they do not realistically utilize the powers given to them. So, even if women are given the participation in local decision making bodies of the village, it is merely for the name sake. They *are* either prohibited by male members from going and attending meetings or do not have any actual say in the decisions taken. Thus, following efforts should be made by *voluntary* organizations for the psychological empowerment of women:

1. Generating awareness regarding women rights and their legal protection.
2. Including skills for institution building and assets' management.
3. Promoting realistic participation in decision making bodies.
4. Generating self confidence and self sufficiency.
5. Assertiveness training.

As most of gender-specific stressors in women are the result of gender biases related with subordination, powerlessness and lack of recognition of work, psychological empowerment of women should be the first step in all efforts towards women's participation. Women should be made to understand their own personal identity, worth and potentials. Their participation will be worthwhile only if they are made to do so. Stress is a matter of cognitive appraisal or perception. When women are psychologically empowered *and* assertive enough to express themselves, their needs, problems and demands and the psychological pressure generated by extreme tolerance, lack of say and suppression of emotions gets released. They feel eased up and can adapt themselves with even greater role burdens without any distress and resulting health outcomes.

Role of Family: Recognizing the family as an important basic social unit, there is a need for strengthening the family. There is a greater need for mutual sharing of responsibilities in home and outside. Empowering women in the family will involve reorganization of career paths and greater economic empowerment.

Muscular Relaxation: Stress is the part of life and we cannot just get away from stress. Training in muscular relaxation act as a wonderful strategy to neutralize the harmful effects of daily hassles and stress. Relaxation is a direct 'physiological opposite to tension or excitement and has potent al advantages. Stress related problems such as hypertension, tension headaches, insomnia etc. may also be eliminated *by* practicing it. Anxiety levels may be significantly reduced. It also reduces the likelihood of the onset of stress related disorders. An important psychological consequence of relaxation is that the individual's level of self esteem and self assuredness is likely to be increased as a result of improved control over stress reactions. By *now,* most of the psychologists are working 'in the urban areas but there is a great need of psychologists *in rural* areas too. Governmental and non-governmental organizations should seriously take into consideration, the inclusion of *psychologists* in their working team so that the psychological empowerment of women and psychological health could be promoted.

Conclusion

Women's roles and relations with environment are vital and crucial, especially India where they are the sole providers of water, fuel, fodder food and other basic necessities. The modern concept of development contributed to an increase in economic and in qualities and growth of poverty, notwithstanding its promise improved living conditions. Degradation of resources base interferes with women's domestic responsibilities affecting them psychologically and physically.

References

C.S.E. (1985), The State of India's Environment, 1984-85, The Second Citizens Report, New Delhi: Centre for Science and Environment.

Lazarus, R.S., (1966) Psychological Stress and Coping Process, New York: McGraw-Hill.

Satyavathi, G.V. (1993), Stress and Women's Health, A Report of World Health Organization, Geneva.

TERI (1992), Human Exposure to CO and TSP due to Bio fuel Combustion, New Delhi: Tata Energy Research Institute.

WHO (1993), Psychological and Mental Health Aspects of Women's Health, A Report of World Health Organization, Geneva.

WHO (1991), Epidemiological, Social and Technical Aspects of Indoor Air Pollution from Biomass Fuel, Report of a WHO Consultation, Geneva.

7

Environmental Education in India

A. Srivasacharyulu and K. Hari Babu

Education is essential for generating widespread awareness on environmental problems. Mass media plays significant role in creation of awareness for plan of action. Without proper education, efforts made by media and other organizations in awareness analysis-action chain does not move smoothly and effectively. Environmental education is the process of recognizing values and clarifying concepts in order to develop skills and attitude necessary to understand surroundings. Environmental education also entails practice in decision making and self formulating a code of behaviour about issues pertaining environmental quality.

According to UNESCO and UNEP, environmental education is a sustained process in which the individuals gain awareness of their environment and acquire the knowledge and skills to enable them to act individually as well as collectively to solve future environmental problems. It is evident that environmental education has four components awareness, values or quality, concentration and sustainable development. In fact, lack of awareness has led man to be harsh to nature (Mazhar Ali Sabri, 2004:7).

Though the history of human beings is about 500 to 600 thousand years old and since his very inception on the planet, man has been modifying the environment. It is during the last three centuries and particularly since the Industrial Revolution in which man has made galloping progress, and this exploitation of nature has led us to a crisis. In this period man has plundered the natural treasures with the help of modern technology. But this exploitation of nature has threatened the very existence of human being on this planet Earth.

In the recent past, industrial (Bhopal gas tragedy, India), nuclear (Chernobyl radiation catastrophe, USSR) and maritime disaster (oil leakage on coast of France and Alaska) have forced the scientists, sociologists and independent research workers to realize the seriousness of the impending catastrophe. The revelation of hole in the ozone layer over Antarctica, Green-house effect, destruction of rain-forest, poverty and population are the problems of greater magnitude for homosapiens.

The population explosion and scientific and technical developments with callous and inefficient management practices are leading to depletion of natural resources. Thus, success in natural resource conservation depends upon a thorough understanding by people of the values of forests soil, water, wild life and related resources to individuals and to their nation.

Naturally, everywhere the concern is only for the physical environment. But, as we have explained in earlier lines, the word "environment" has such wide application to every kind of milieu that it also connotes a cultural, a social and even an economic setting. It would be pedantic, therefore, to argue that our government should specify its concern only with the physical surroundings, with the air, the land and the waters, the flora and the fauna. To limit thinking, leave alone action, only to immediate health hazards like sources of pollution and the planting of trees, is a poor and partial comprehension of the physical environment which in fact is the country, this India that is Bharat.

This underlines the need of environmental education. In fact, since time immemorial environment has been man's permanent teacher and is so even today. Mankind would never have got anywhere without the vital knowledge about environmental phenomenon. Thus, an understanding of the environment is indispensable for its rational management. This could enable us to predict better the interrelated efforts of some of the challenges facing the world today, which may include demographic changes, availability of food, fodder, energy and raw materials, economic development and utilization of new technology, etc. Better understanding of environment can only be achieved through environmental education. That is why environmental education is so important and essential at present. So, the global need of environmental education as a measure for conservation and pollution-free environment for sustenance of life and healthy living is felt (Monga G.S, 2001:86, 87).

Environmental education deals with man's relationship with his natural and man-made surroundings. It also deals with the dynamics of the physical, biological, socio-economic, political and technological dimensions. It can be seen from two different perspectives, one which treats it as a separate new discipline and the other which treats it as a new dimension to existing curricula cutting across different disciplines.

The aim of environmental education programmes should be, thus, to increase awareness of the environment and its problems; basic knowledge and understanding of the environment and its interrelationship with man; social values and attitudes which are in harmony with environmental quality; skills to solve environmental problems; and appropriate actions to solve environmental problems to create a sustainable environment.

Environmental education needs to be given in local context. India is a land of physical, ecological, social and cultural diversity. Bound by the Himalayan mountain ranges in the North and the Indian Ocean in the South, it has a multitude of climates, soil types and geographical areas and consequently of habitats and wildlife, and accordingly, the content of the environment education should be aligned. Students should be taught about the environment that is visible in their areas.

Environmental education has an important role to play in the promotion of environmental awareness. The knowledge base of a society is one important aspect of its capacity to address and cope with environmental issues. Environmental education is the first step in enhancing this knowledge base. A look at the existing state of environmental awareness and education indicates that the picture is at first glance quite positive, at least in most countries of the developed world. In developing countries, the picture is more mixed, though environmental education has made some inroads. Surveys in the developed world reveal that most people consider themselves environmentalists; the figure is especially high among youth (Chitrabhanu, T.K, 2007:1, 2).

Nature and Objectives of Environmental Education

Though no single precise and universally acceptable definition has emerged as yet, environmental education is broadly a programme dealing with the dissemination of knowledge about environment and the impact of human beings upon it. The environment here includes animals and plants and their ecological systems which are closely bound to the livelihood of people. It enables people to enjoy good health and high quality of life.

The basic aim of environmental education is to succeed in making individuals and communities understand the complex nature of natural and the built environments resulting from the interaction of their biological, physical, social, economic and cultural aspects, and acquire the knowledge, values, attitudes and practical skill to participate in a responsible and effective way in anticipating and solving social problems, and in the management of the quality of environment. "A further basic aim of environmental education is clearly to show the economic, political and ecological interdependence of the modern world, in which the decisions and actions by different countries can have international repercussions. Environmental education, in this regard, helps to develop a sense of responsibility and solidarity among countries and regions as the foundation for new international order which will guarantee the conservation and improvement of environment."

Goals of Environmental Education in India

The Environmental Education Programme in India is based on the following goals: to improve the quality of life; to create an awareness among people to environmental problems and conservations; to help citizens acquire the necessary knowledge for the importance of environmental protection; to develop necessary skills to solve the environmental problems; to create an ability in people to evaluate the different strategies for development in terms of social, political, cultural and educational point of view; to create the necessary atmosphere to make it possible for citizens to participate in decision-making when it concerns the environment; to provide an awareness of economic, political and ecological interdependence.

A deeper analysis shows poor response of major population of the country to environmental education. There are certain problems which are faced by the people such as:

1. Deficiency of environmental education knowledge among enforcing authorities.
2. Lack of responsibility due to ecological illiteracy on part of community.
3. Lack of provision for training of professional manpower in the field of environmental education.
4. Non-availability of information on environmental issues.
5. Lack of coordination among the administrative institutions and educational institutions on environmental activities (Monga, G.S. 2001:87-89).

The fight against environmental degradation is not the only concern of the government; it has to be an issue of every body's concern. 'Silent valley project' in Kerala and the 'Chipko movement' in the Himalayas in India is a welcome sign of people's awareness. The role of rural women folk in the Chipko movement is heartening. Environmental education to understand the intimate relationship between the quality of environment and human well being has an important role to play in the direction of preserving the environment.

Human Impact

Rapid growth of world's population is affecting the earth's surface at ever increasing rate. The early agronomists burned large areas of land to create farm land or pasture, they modified the soil by ploughing, altered the drainage by irrigation, introduced or bred new animals and crops and altered the new natural vegetation structure of many regions.

Renewable resources are being consumed at rates that far exceed the speed at which they can be regenerated or replenished. A hectare of land can be destroyed within an hour but it may take several decades for the forest to regenerate itself.

Water pollution is the most serious environmental problem facing developing countries because of its direct effect on human welfare and economic growth. The diversion of fresh water to supply agricultural, industrial, domestic and municipal needs stretches hydrological cycle. Fertilizers and consumer goods also have a powerful impact on water quality.

Role of Media in Environmental Education

The network of media and information technology is vital to disseminate environmental education in motivating people. Technology acts as a practical tool and media acts as a communication tool. Information system on environment must be considered as one of the management tools for planners and policy makers. The availability of adequate information is a must to improve project implementation and to utilize the available resources in a sustainable manner.

Mass media like radio, television, newspapers and journals must be used to educate the people to understand the forces of nature as a whole and their participation in the process to understand and utilize rather than to destroy due to lack of understanding of environment. Posters, brochures, banners, exhibitions, photo shows are still found to be effective in dissemination of message. Labels and slogans that can be pasted on the bags, boxes, packets, are other simple ways for disseminating information and form the public opinion.

Environmental problems show a reasonable correlation between general awareness and the news coverage. It is a fact that in this decade there is an increasing number of books and magazines and T.V. channels dealing with environmental issues but they are unable to reach up to grassroot level. In many cases, the public responds as a result of increased awareness of the dimensions of the problems due to the catalytic role of the mass media.

People's participation in environmental plan of action has great potential. Participation of the community can be ensured by organizing fairs and festivals like *van mahotsava, varsha vandana* and celebrating world environment day etc. Thus, access to media and other modes of communication is primordial for creating awareness about environment.

Role of NGOs in Environmental Education

It has been realized that the government system has failed to deal with the environmental problems. If any serious effort is being made, it is by the courts. From district courts to Supreme Court all are taking on environment related cases. Over the past two decades NGOs are actively engaged in implementing various environmental protection and development programmes to mobilize communities. It is felt that NGOs with their grassroot level knowledge have an advantage in organizing the initial training programmes and demonstrations for people-led programmes as envisaged by the participatory rural appraisal (PRA) techniques. PRA techniques stress upon the importance of listening and learning from grassroots to form the basis of the planning and programme implementation.

UN conferences on environment help to focus world attention on the dangers to human survival and quality of life posed by the continuous degradation of basic ecological assets. The summit meeting at Colorado of 200 NGOs from 55 developing countries on "Development during the next coming Decades" and other meetings have reiterated the positive role played by NGOs in the implementation of these programmes via a bottom-up approach and many to many fashion rather than top to bottom approach.

Recently, the importance of ecological and economic forces at the grass root level has been recognized. UN Environment Programme and the various environment bodies throughout the world have come up for the implementation of various projects on environment. It is encouraging to see this being emphasized and given due importance at the highest levels. However, the thrust of planning from above will take its own time to give way to planning from bottom and implementation by the people themselves at micro level in order to appreciate the ecological and economic interdependency that is harmonious and sustainable.

Relationship between Development and Environment

Environmental education offers the opportunity to close the gap between rhetoric and reality, between saying we should follow sustainable development paths and actually taking the steps to persuade societies and individual to do so.

Protection of environment is an essential part of development. Without adequate environmental protection, development is undermined: without development, resources will be inadequate for required investment and environmental protection will deplete. This growth brings with it the risk of appalling environmental damage. However, it could bring with it better environmental protection, cleaner air and water and the virtual eradication of acute poverty.

Damage to environment affects present and future human welfare. It harms human health, reduces economic productivity, and leads to the loss of amenities' (a term that describes the many other ways in which people benefit from the existence of an unspoiled environment). Some problems are associated with the lack of economic development, inadequate sanitation and clean water, indoor air pollution from biomass burning. Many types of land degradation in developing countries have poverty as their root cause (Mazhar All Sabri, 2004: 9).

Development and environment explores two-way relationship. First, environmental quality—safe, plentiful and healthy water and air—is itself a part of the development efforts. If the benefits from rising incomes are offset by the costs imposed on health and the quality of life by pollution this can not be called development. Second, environmental damage can undermine future productivity. Soils that degraded, aquifers that are depleted and ecosystems that are destroyed in the name of raising income today can jeopardize the prospects for earning income tomorrow.

Low agricultural productivity caused mainly by poor incentive and poor provisions of services has delayed the demographic transition and encouraged land degradation and deforestation which, in turn, lowered productivity. Africa's forest declined by 8 percent in 1980; 80 percent of Africa's pastures and range areas show signs of damage; and in such countries as Burundi, Kenya, Lesotho, Liberia, Mauritania and Rwanda, fallow periods are often insufficient to restore soil fertility.

There are two sets of policies for development and environment: first, that seeks to harness the positive links between development and environment by preventing or correcting policy failures, improving access to resources and technologies and promoting equitable income growth and second, regulation and incentives which are required to force recognition of environmental values in decision making (Sabri, M.A 2000:29-34).

The fragile environment is undergoing unsustainable levels of stress from growing populations, increasing demand of resources and pollution from household, agricultural and industrial sectors causing additional pressure on land and water, increasing air and water pollution and increasing of solid waste. The world commission on environment and development warned that the world faced unacceptable levels of environmental damage. The size of earth is fixed while its resources are being depleted abruptly. Environmental Impact Assessment (EIA) is 'the evaluation of the effects likely to raise from a major project (or other action) significantly, affecting the natural and man-male environment (Wood C., 1995: 337). It was emphasized that EIA should be incorporated into all existing and upcoming development projects. EIA is thus to prepare necessary action plan to present, eliminate or mitigate the adverse impacts of any developmental activity as a part of the overall environmental management plan.

Environmental Degradation

With the world shrinking, came the realization that environmental degradation in more countries affected others and the problem had to be tackled on a global scale—many species of fauna and flora are extinct. As dangerous as the threat of extinction of some species is the threat of changes in the atmosphere. Gases have polluted the air we breathe and are destroying the ozone layer. The green house effect has raised world temperature. Glaciers have begun to melt as has ice in the arctic zones. Levels of oceans are rising and if it continues to do so many low lying countries will submerge under water. There is, therefore, a need for tailoring the pace and pattern of economic growth to the earth's carrying capacity. People have the potential or capacity to make environmental and development sustainable.

Depletion of Wetlands: Millions of species of plants, animals and other organisms enrich our environment. Awareness of the importance of this biological diversity has grown in recent years along with concern that more effective action is needed to preserve it. There is an urgency because destruction of ecosystems and species extinction entail irreversible losses. Ecological degradation of wetlands together with pollution has resulted in the loss of flora and fauna. The high amount of fertilizers and other inputs required in agriculture for increasing the productivity has led to the degradation of the environment. Coral reefs and mangroves are threatened by increasing discharge from industrial establishments along the coastal belt. The mangroves of Sunderbans delta have been reduced to half (Kumar, A. Biju, 1999:10-15). Organic pollution of water is the most serious problem in most of the developing countries. There has been a marked increase in the surface water pollution caused by the use of fertilizers, pesticides and acid rain (Mazhar Ali, Sabri D., 2004:10).

Deforestation and Desertification: Over the past several centuries the World's forests have declined one fifth, from 5 to 4 billion ha. When forests are cleared their capacity to withhold carbon dioxide from the atmosphere is lost. Deforestation in rainforest and moist deciduous forest areas is of global significance because it affects regions rich in biodiversity. The removal of forests in dry areas depletes already scarce resources such as fuel wood and fodder. Evidence from the Sahel region of Africa suggests that the availability of forest resources is critical in determining the carrying capacity of its agricultural and pastoral communities.

FAOs findings indicate that annual deforestation rates increased from 2.0 million hectares during 1976-81 to 39 million ha. in 1981-90. Asia and the Pacific region have the fastest rate of deforestation and species extinction among the tropical regions of the world. Although India's forest cover area is not currently shrinking but forest degradation continues due to urban demand. India's forests provide an estimated 41 million cubic meters of firewood per year, yet current annual demand is thought to the 240 million cubic meters. Another major pressure on forest resources is the increasing demand for timber and paper.

Overexploitation and illegal cutting of indigenous forests

is a matter of great concern. The lack of appropriate technology for maximum utilization of the raw materials, low recovery, poor silvicultural practices, low, budgetary provision for recurrent forest operations have all hampered effective management. As a result fuel wood, timber, pole wood and carving wood are being exploited at unsustainable levels (Sabri, M.A 1999: 37-39). Arid and semi-arid lands are suffering from an increased rate of desertification and frequent drought.

Excessive use of chemical Fertilizers: Excessive use of chemical fertilizers may decrease soil productivity in the long run. Chemical fertilizers lead to soil acidification and also deplete organic matter. As a result, the volume of soil pores and soil micro-organisms are reduced, and thus, compaction, reduction in water holding capacity and soil erosion occurs. Nitrogen in the form of water soluble nitrates is the most common and problematic contaminant from chemical fertilizers, although phosphorus and potassium also pollute aquifers. It was not possible to raise food production without using chemical fertilizers. These chemicals, in turn, cause environmental hazards resulting in a vicious cycle.

Fertilizers are used in about 62 percent of India's farmlands and about 43 percent of rain-fed land is fertilized. In part, this can be attributed to the fact that soil testing facilities are not widely used to advise farmers on fertilizers needs. Such testing could boost yields and efficiency by eliminating nutrient deficiencies and unnecessary fertilizer application.

Excessive use of Ground Water and other Natural Resources: Despite decades of warning about pollution and efforts to control it, people are still being exposed to toxic pollutants. Pollution from agricultural land caused by the leaching of nitrogen fertilizers has been detected in the groundwater in many areas, In Haryana, for instance, some well water is reported to have nitrate concentration ranging from 114 mg/litre to 1800 mg/litre far above the national standard (45 mg/litre).

Policy Implications

Developing countries should seek to achieve sustainable consumption patterns in the development process, ensuring basic needs for poor people. Governments should monitor and implement policies to reduce pollution and environmental degradation and to safeguard the natural system that supports renewable resources as well. Policy options exist for reducing pollution for preventing the exhaustion of resources and for shifting resource consumption to more sustainable patterns. New companies must go through an elaborate clearance process in some cases including Environmental Impact Assessment (EIA), before they begin operations. Action should be taken for achieving a sustainable future.

Environmental education should deal with the underlying relationship between environment and development for a healthy society. Our development must be designed in such a way that it does not damage the environment but builds and restores its health. To promote environmental education, both formal and non- formal, there should be involvement of each and every human being. Apart from governments, people's organizations including NGOs, cooperatives, self-help groups (SHGs), mass media and local institutions should be involved to promote environmental ameliorations by using Environmental Impact Assessment (EIA) and Participatory Rural Appraisal (PRA) techniques.

Environmental degradation has caused concern at international, national and local levels. This, in turn, has led to the emergence of the concept of sustainability of communities, ecosystems and projects. Achieving sustainability must become a control objective. The pollution of air, water and soil by agricultural, industrial and human wastes has been increasing considerably. There is, therefore, a powerful need to implement the four R's Action Plan: Refuse, Reduce, Reuse and Recycle.

References

Mazhar Ali Sabri (2004), "Environmental Education-Some Key

Issues", *Kurukshetra,* Vol. 52, No. 8, pp. 7-10.

Chitrabhanu T.K. (2007), "Introduction to Environmental Education" in, Chitrabhanu T.K. (ed.), *Environmental Education*, New Delhi, Authors Press, pp. 1-2.

Monga G.S. (2001), "Place and Problems of Environmental Education in India" in Monga G.S. (ed.), *Environment and Development,* New Delhi, Deep & Deep Publications, pp. 86-89.

Kumar, A. Biju, (1999), *Science Reporter,* December, 1999, pp. 10-15.

Sabri, M.A., (1999), "Balancing Development and Ecology", *Social Welfare,* New Delhi, Vol. 46, No. 3, June 1999, pp. 37-39.

Sabri, M.A., (2000), "Poverty and Food Security: Problems and Prospects", *Kurukshetra,* (A Journal of Rural Development), Ministry of Rural Development, New Delhi, Vol. 49, No. 3, December 2000, pp. 29-34.

Wood C, (1995), "Environment Impact Assessment: A Comparative Review", Harlow: Longman, p. 337.

8

Environment and Development

M. Galaiah

Environment is that set of surroundings which influences the life and activities of an animate object, human or animals. It consists of the totality of all those external conditions and influences which affect the life and development of a living being. As P. Gisbert says, "Environment is anything immediately surrounding an object and exerting a direct influence on it". It is of two types:

1. Natural and physical or geographical environment, and
2. Man-made (social or technological) environment otherwise called as material culture.

In so far as natural environment is concerned, generally speaking, the agro, pastoral civilisation respects the landscape and strives to adopt itself to it while industrialisation demolishes it and transforms it into a threat to life completely. It remodels the appearance of the earth almost entirely breaking up and removing the surface of the soil with quarries or mines or burying it under slag, levelling it by means of viaducts, covering it with large cities even artificially to revive the original landscape through parks and gardens. Large cities spread and unite, forming urban regions and no connection with the natural region any longer, exists.

With the resurgence of interest in protecting the environment on a wide scale and people's movements all over the world including India over the adverse impact of development projects, environmental issues cannot any longer be brushed aside with contempt by politicians and economists. There was a time, particularly during the Stockholm Conference on Environment and Development in 1971, when it was fashionable to argue that development is the primary concern of developing countries who have to fight poverty and unemployment on an urgent basis, and that environment, therefore, should have low priority. Much polluted water has flown down the Ganga since then, and perceptions have been modified if not changed completely.

Development continues to be of primary concern for that is the only way we can overcome the problems of poverty and unemployment in developing countries. But it is now recognized widely that development without regard to environmental consequences cannot be sustained and defeats the very purpose of development by aggravating poverty and unemployment. Environment is not just a matter of aesthetic concern or of saving some white legged mouse from going extinct as in a popular cartoon serial which lampoons environmentalists. Environmental concern arises essentially on two grounds.

First, a reckless use of resources to maximise short-run gain can lead to depletion which would adversely affect development in future, which may not be a very distant future. It is not in the interests of sustainable development. Second is that one does not have to wait for future generations to feel the adverse impact on environment. Livelihoods are lost, sources of drinking water are made unusable, and victims are subjected to agonising misery (some times irreversibly) due to negative externalities of so-called development projects. While a few already better off sections may become still better off, many others become worse off. The former cannot offset the latter's effect.

What is involved thus is not just environment, but human rights, equity and social justice. Neither economists nor politicians can afford to be indifferent to these issues. No economist worth his salt would advocate today stepping up the rate of growth of production of commodities for its own sake. Growth with justice became a slogan long back in the 1960s when environment issues had not come to the fore. We find today that environmental issues involve essentially issues of welfare and justice concerning the present and future generations.

The conflict between welfare, justice and environment on the one hand and development on the other becomes much less acute if we consciously avoid the confusion between development and growth. I quote Herman Daly: "By growth' I mean quantitative increase in the scale of physical dimensions of the economy; i.e., the rate of flow of matter and energy through the economy *(from* the environment as raw material and back to the environment as waste), and the stock of human bodies and artefacts. By 'development' I mean the qualitative improvement in the structure, design and composition of physical stocks and flows, that results from greater knowledge, both of technique and *of purpose.* Simply put, growth is quantitative increase in physical dimensions; development is qualitative improvement in non-physical characteristics. "An economy can, therefore, develop without growing, just as the planet Earth has developed (evolved) without growing" (emphasis in original) (Daly, 1992: 36).

Though, admittedly, economic development has accompanied economic growth since the Industrial Revolution, one can not only distinguish between the two, but also reach higher levels of development without corresponding rates of growth. The fact that we felt the necessity to have a Human Development Index (HDI) separately from GNP is an eloquent testimony for this. Several developing countries have reached higher levels of development in spite of lower levels of GNP. The pertinent point here is that while our obsession with economic growth can be environmentally damaging, economic development need not have such an effect. The shift of focus from commodities to capabilities, borrowing Amartya Sen's words, is a good news for both development enthusiasts and environmentalists.

It is true that it is not possible to avoid economic growth in terms of physical quantities of commodities while aiming at economic development in qualitative terms, particularly in a developing country like India, though developed countries are in a much better position to do so. To achieve a better quality of life, we also need to produce more in developing countries. Even when we concede this, we have to look more and more for opportunities that improve the quality of our life without having to spend corresponding amount of material inputs. T.W. Schultz recognised long back that even economic growth or agricultural production and income to be more specific, could be achieved through greater efficiency and knowledge without corresponding expenditure on material inputs (Schultz, 1960 and 1961).

At higher levels of economic growth, the scope for such improvement in efficiency should increase rather than decrease. On the contrary, capital-output ratios have tended to increase over the decades. This cannot be brushed aside as inevitable, reflecting the law of diminishing returns. This Law need not operate at the aggregate level of the economy as a whole, particularly when improvement in efficiency in the use of capital and inputs is allowed. An increase in capital-output ratio occurs due to quite different factors, the most important of which is the vested interest of politicians, bureaucrats and engineers in keeping investment levels high.

Increasing investment levels may not be bad, but doing it wastefully is. Roads are built at enormous cost which do not last a single monsoon. Costly bridges are built which do not last even a decade. Even manufacturers are not interested in manufacturing sturdy and long lasting goods, as that would decrease replacement demand. The cherished values are such that almost all economic agents seem interested only in maximising short-term gains. This is most disturbing to environmentalists. Sustainable development cannot be achieved with such an economic behaviour dominating the economy. A throw-away economy is not a sustainable economy.

Improving the efficiency of investment and enhancing the

quality of life in ways that minimises the use of material inputs constitute very important strategies of reconciling economic development with environmental concerns. This tends to reduce the demand for energy and other natural resources which are scarce. Environmentalists and environmental economists have always emphasised the demand side and the need to restrain it, because it is not within the means of this planet Earth to satisfy growing and wasteful demand forever. While admittedly we need more water and energy per capita in India, it should be possible to improve the efficiency of their use and moderate the demand for it. Environmental economists have indicated that one of the ways of restraining the demand for natural resources is to price them in such a way that prices reflect their true scarcity, or at least the social cost of supplying the resources including the cost that we impose on future generations by their present use. While this can be a useful strategy, there are some problems with it.

The most important problem with using price as the regulator of demand for natural resources is that where the demand for it is price inelastic, demand is hardly restrained. This is so particularly in a highly inegalitarian society, where the rich indulge in luxuries they want and hardly care for prices of such things as water and energy. On the other hand, such a pricing can place an unduly large burden on the poor. Non-price or physical regulation may appear clumsy particularly in the context of economic reforms, but it seems one cannot do without them. This is not to say the price policy is irrelevant or always ineffective. It can work to a significant extent. Had it not been for the hike in petrol prices, more fuel-efficient cars would not have been designed.

However, even this has limits. The hike in petrol prices has hardly moderated the total demand for petrol, it has hardly reduced the number of cars, and it has hardly led to a replacement of energy-inefficient private transport by public transport. Economic instruments of environment policy work no doubt, but they have serious limits. We cannot escape from overall planning and regulatory instruments which may, however, have to be used in combustion with economic instruments for making them more effective.

We have referred above to two ways in which environmental probiems arise in the course of economic growth—depletion and negative externalities of development projects. While a concern for achieving economy and efficiency can help in taking care of depletion, a concern for achieving equity is necessary in tackling negative externalities. Development projects have to incorporate as their essential part a provision for compensation for those who are displaced or adversely affected by them and for their complete rehabilitation and resettlement. In the past, this aspect was given hardly any attention, with the result that even while some sections benefited from development projects, others actually became worse off and impoverished. Thanks to people's movements like those led by Medha Patkar, both, the Government of India and State Governments have started giving more serious attention to this problem.

But the problem is still formidable, if not impossible. Displaced people are offered compensation for lost lands on the basis of prices registered in sale deeds in the recent past, or as multiple of average value of yield from the acquired land. In either case, the displaced persons are not in a position to purchase the same quality and quantity of land elsewhere with the compensation money. Sale deeds usually under record prices to reduce stamp duty and give no indication or real market price. Even the capitalised value derived from yield falls far short of market prices that the displaced persons have to pay for purchase. This is particularly so where not enough land is available for purchase, and land prices are pushed up on the expectation of demand from the displaced persons. *Karnataka* is now trying out a new approach called 'consent price', which is actually a price negotiated between the affected person and the government, with the option of approaching a special court in case negotiation fails. If the negotiation succeeds and consent price is settled, the person does not have the option of going to the court to settle the price. The consent prices are usually higher than *what* the courts are expected to award on the basis of capitalization method. In spite of this, it is reported that the displaced persons find it difficult to purchase new land with the money obtained. Since a land to land policy is not feasible the government is trying other means also, by providing non-farm employment opportunities to the displaced and/or affected families but the amount of such employment created seems to be much smaller than what is required.

The money required, for real resettlement and rehabilitation of displaced or affected people to make them no worse off than before, is likely to be enormous. Project designers have often the temptation to underestimate these costs which can do grave harm to the project through subsequent cost escalation. It is more prudent to fully estimate and provide for this and incorporate it as a part of the development project in judging its worthwhile ness. This can lead to formulation of alternative projects so that a choice can be made on an informed basis in favour of a project that minimises the total cost in relation to expected benefits.

Though the general perception is one of conflict between environment and economic development, *in* quite a few ways even economic growth is helpful both in improving environment and in becoming less vulnerable to harsh effects of environment. Insofar as *economic* growth and development creates alternative avenues of employment, it can reduce pressure on land and forests and can prevent land degradation and deforestation. One of the major reasons, for loss of forest cover in developing countries including India, is conversion of forest land into cultivated lands since employment opportunities in the non-agricultural sector are limited. A boost to employment in the non- farm sector can thus help environment. Economic growth and development also makes available resources with which it is possible to allocate more funds for improvement of environment. Economic development also creates greater environmental concern and raises the demand for environment preservation as a source of recreation. Even the estimates of existence value of environment such as wildlife, improves with the economic development. Economic development, by making people less dependent directly on environment, also makes them less vulnerable to environmental uncertainties and natural vagaries. It was found through a cross-section study of districts in southern India at two points of time (1971 and 1981) that the differences in the quality of life between drought prone and non-drought prone areas have declined in the process of development *(Nadkarni,* 1985, esp. Chs. 2 and 6). Developed countries in Europe and America are able to carry on and prosper in spite of harsh weather.

There is a wrong impression prevailing even among environment interested social scientists that problems of industrial and urban pollution are not as important in India as in developed countries. It is no doubt true that problems of land degradation and deforestation are more serious in developing countries like India than in the developed. But this is not to say that therefore problems of pollution can be ignored. Far from it, India is fast industrialising, and has emerged as a major industrial power. With it also have emerged serious problems of pollution. But we are not doing enough to tackle it. In a recently completed comparative study undertaken by me in collaboration with the Institute for Environmental Studies, Amsterdam, we found that the standards for pollution treatment in the industries selected for our study were lax in India compared to Netherlands.

What is more, even the parameters laid down in the standards are very inadequately monitored in India, let alone enforced (Nadkarni et al., 1995). The air in our city centres has become unbreathable, and water pollution by industries is making drinking water scarce in many places. By pursuing 'development' in this way, we are only pushing up health care costs both for the victims and the government. Just because industries do not bear them, we cannot pretend that they are not there, nor can we pretend that even if they are here, they are a price to pay for poverty alleviation.

The fear that a stricter regime of industrial and urban pollution control will adversely affect industrialisation and employment is unfounded. The so-called conflict between 'development' and environment in this regard is unnecessarily exaggerated. From a study of selected firms in Karnataka, it was found that the costs of pollution control were well within affordable limits. Thus the annualised cost of pollution control in these firms varied from 0.2 to 1.9 percent of annual turnover (Nadkarni et al., 1995). A stricter implementation of pollution control could, of course, push up these costs, but they can still be considered absorbable. The series of hikes in petrol prices since 1973 have been far sharper in comparison with the extra burden imposed by pollution control. The hikes in petrol prices have hardly reversed industrialisation or employment generation.

Nonetheless, we have to recognise that the costs of pollution control and prevention imposed on the developing countries can be considerable. The developed countries of today had the luxury of ignoring pollution control when they were at comparative levels of per capita income of the present developing countries. They chose to take care of pollution problems only when they reached high levels of income which are still beyond the reach of developed countries.

They have also tried to reduce pollution by simply exporting polluting industries to the developing countries, a route which is not open to the latter. On top of this, the developed countries are expected to conserve forests in global interests, while the human and material cost of such conservation is to be borne by them. Obviously, this is a situation which is blatantly unfair to developing countries. They need significant help from the developed countries in solving environmental problems The mechanisms that the international community have developed so far for this task are utterly inadequate. With some real help from the developed countries, the developing countries can more easily reconcile economic development with environment. But they can help in making it environmentally sound and sustainable in global interest.

References

Daly, Herman (1992): "The Economic Growth Debate: What some economists have learnt but many have not", in Markandya and Richardson (eds.): The Earth Scan Reader in Environmental Economics, London; Earth scan, pp. 36-49.

Nadkarni, M.V. (1985): Socio-economic Conditions in Drought Prone Area, New Delhi: Concept Publishing.

Nadkarni, M.V., G.S. Sastry, Onno Kuik, Frans Oosterhuis and Ackerman (1995): Best of Both Worlds: A Comparative Assessment of Environment Policy Approaches in India and the Netherlands, IDPAD Study, awaiting publication.

Schultz, T.W. (1960): "Capital Formation by Education", Journal of Political Economy, Vol. 68 (6), December.

Vallaunx Comille: Encyclopaedia of Social Science.

9

School Education for Environment

N. Alivelu Manga and G. Rajaiah

"If you wish to plan for a year, sow seeds;
If you wish to plan for 10 years, plant trees;
If you wish to plan for life time, develop men."
—*Kuang Chung Trim*

A healthy natural environment is a fundamental prerequisite for sustainable human development and human survival. Environmental information is a key element in achieving a good level of public involvement and participation in the process of sustainable development. India has enormous resources in its natural biodiversity and traditional knowledge systems that have the potential to be harnessed for sustainable economic development. Effective management of these resources calls for a change in the attitude of the public and the civil society in order to identify, assess and record these resources.

There is a need to promote environmental education and awareness to educate and inform all stakeholders and the public that irrational depletion of national natural resources is destroying the basis of prosperity for future generations and that as forests disappear, land becoming infertile and water is exhausted or polluted, it is the poor of today, especially children and women, who suffer the most. There is an urgent need to improve public awareness and understanding of environmental issues with a view to promote the conservation and wise use of natural resources at the community level.

The Government of India recently introduced environmental education throughout the country among the school children. The respective teachers who are working in high schools are to implement this programme. In this paper, we have tried to examine the implementation of environmental educational programmes in high schools of Andhra Pradesh by choosing Warangal district for our study. In order to know the present status and implementation part of environmental education at the school curriculum, a sample survey was conducted in Hanamkonda mandal of Warangal district. About 20 schools were surveyed—rural and urban, private and government. The main purpose of the survey was to examine its implementation part both in private and government schools.

After having surveyed among schools, it is noticed that the teachers working in primary and secondary schools are showing less interest about environment and its education imparting to student community. The study revealed that about 80 percent of the teachers working in these schools are not implementing programmes pertaining to environmental education. It is rather pity that the head of the institutions are also not feeling any responsibility to implement the programme effectively. We feel that this type of activity is needed and effective implementation of environmental education among school children is a must for future generations also. In this regard, teachers are to be trained properly. Supervision of implementation of this scheme also needs urgent requirement from the Government side. We shall discuss below how the environmental problems emerge and which are the problems that need to be solved and attended to.

Nature of Environment and its Problems

Environmental problems have emerged due to pollution and over exploitation of natural resources. Population explosion is the prime cause of many such problems. Environmental problems affect human health and quality of life. Environmental pollution damages the natural environment which affects vegetation, animals, crops, soil and water. The first report of the British Royal Commission on environmental pollution (1971) said, "The best insurance for the environment is a commitment on behalf of the public to prevent the deterioration for air, water and land."

The amount of atmospheric CO_2 apparently remained stable for centuries at about 260 parts per million (ppm), but over the past 100 years it has increased to 360 ppm. The significance of this change is its potential for raising the temperature of the Earth through the process known as the "green house effect". Studies show that the ozone layer is being damaged by the increasing use of industrial chemicals called chloro fluoro carbons (CFCs) which are used in refrigeration, air conditioning, cleaning solvents, packing materials and aerosol sprays. CFCs can remain in the atmosphere for more than 100 years, hence ozone destruction will continue to pose a threat for decades to come.

India is now the world's fifth largest fossil-fuel CO_2-emitting country: the emissions having grown at 6 percentage points a year since 1950. There have been several studies of the impact of present global warming on India, especially on food production and in coastal areas. A study by the Jawaharlal Nehru Technological University in 1993 found that a one-meter rise of water level in the sea would inundate approximately 5,800 square kilometres of coastal area and directly affect 70 lakh people; the economic loss for the nation would stand at ₹ 2,30,300 crore and for Mumbai at ₹ 400 crore at present prices. A UNEP (United Nations Environmental Programme) team that went to the Himalayas recently, found that a glacier near the first camp that Edmund Hillary and Tenzing Norgay set up during their conquest of the Everest in 1953 had receded by 5 km. and that a series of small ponds had now formed into a big lake.

The Indian Constitution laid down the responsibility of Government to protect and improve the environment and made it a "fundamental duty of every citizen to protect and improve the natural environment including forests, lakes, rivers and wildlife", but it is only on papers. In practice, it is not implemented properly.

In order to address environmental issues, various programmes have been chalked out by the Government of India. Environmental awareness cannot be addressed adequately through the formal education system. Awareness must also be created through non-formal and formal education methods, as the national overall literacy rate is still very low. The central government as well as states together can promote environmental education under various schemes. Educational policies of the government, which are relevant to the promotion of environmental education among the school children, are listed below for all the students in the country. If these policies are followed by each and every student, who are going to be future citizens, the environmental problems mostly are arrested and solved. The policy programmes are explained below.

A. National Policy on Education (NPE), 1986: This is an important development in the policy towards education of environment, which specifies, among other things, the following; The National Policy on Education (NPE), 1986 states that the "protection of the environment" is a value which along with certain other values must form an integral part of the curriculum at all stages of education. The policy states that "there is a paramount need to create a consciousness of the environment. It must permeate all ages and all sections of society, beginning with the child. Environmental consciousness should inform teaching in schools and colleges. This aspect will be integrated in the entire educational process". There has been special emphasis on the need to give importance to environmental education, and this has been kept in view while designing curricula, framing the syllabi and developing text books. The philosophy behind the curricula is that environmental education could be infused into the curricula especially at the primary level.

B. National Environmental Awareness Programme

(NEAC): The Ministry of Environment and Forests (MoEF), Government of India started the National Environment Awareness Campaign (NEAC) in 1986 with the aim of creating environmental awareness at all levels of society. Under NEAC, the Ministry provides financial assistance to selected non-governmental organizations, education and training institutes, community organizations, etc. to create massive awareness among citizens of India. Diverse target groups ranging from students/youth/teachers to rural and tribal population, women, professionals and the general public are covered under this campaign. The Campaign programmes are basically composed of a spectrum of short duration programmes. As part of it, one of the programmes is Eco-club programme in schools.

Eco-clubs in Schools: A non-formal pro-active system of imparting environmental education to school children by involving them in various environmental activities through the scheme of Eco-clubs has been evolved by the Ministry of Environment and Forests. An Eco-club may be set up in a middle/high school and should consist of a minimum of 20 members and a maximum of 50 members, particularly interested in the conservation and protection of the environment, and willing to dedicate time and effort on a regular basis towards this end. The members may be drawn from students belonging to classes from VI to X. Each Eco-club will be under the guidance of an active teacher in the school concerned. The Ministry provides some financial support per annum per Eco-club.

C. Environmental Orientation to School Education (EOSE): Under this, the National Council of Educational Research and Training (NCERT) is implementing the scheme of EOSE. Under the Scheme, financial assistance is provided to create environmental consciousness among the students. These activities will be taken up on project basis. Each project area may also consist of a few blocks/districts having homogeneous ecological conditions.

The NCERT has been entrusted this responsibility in the country for some blocks/districts, by creating the project cells. If an agency undertakes this project, it will set up project cells for each project area to design and organize various educational programmes in the schools, keeping in view the local environmental situation and concerns. However, each project under the Scheme would only cover limited areas. The following guidelines would govern and regulate assistance for taking up innovative programmes for EOSE.

The overall aim of this Scheme is to promote experimentation and innovation, and to complement in diverse ways the goals spelt out in the NPE, 1986 and NCF, 2005 for creating environmental consciousness and related behavioural practices among students.

Some of the activities envisaged under the Scheme are: (i) Encouraging and undertaking curriculum enrichment projects in the area of environment, including making environmental education an integral part of curriculum in school education, leading to development of local-specific teaching-learning materials (e.g. brochures, posters, maps, charts, art and artefacts, models, audio and video materials as well as CDs and websites). In addition to that, organization of exhibitions, literary gatherings, dramas, debates and discussions, dances, film shows, street-plays, *melas* and other such activities including those which the panchayats may suggest can also be undertaken. (ii) Action research/experimental/innovative activities, including activities aimed at generating good primary data on local environmental parameters wherever necessary involving the panchayats. The Scheme is good, but there is a lacuna in implementation both at the national and the state level.

D. National Green Corps (Eco-clubs): The best way to attempt to bring about a change in the attitudes in the society is through children. They are the single most important influence in any family. With this realization, the Ministry of Environment and Forests, Government of India has decided to launch the National Green Corps Programme (NGC) in all districts of our country. It appears that even here also involvement of children in this Scheme was not found and it remains on paper only. The main objectives of this programme is to educate children about their immediate environment and impart knowledge about the eco-systems, their inter-dependence and their need for survival, through visits and demonstrations and to mobilize youngsters by instilling in them the spirit of scientific inquiry into environmental problems and involving them in the efforts of environmental preservation. A programme of raising 'National Green Corps' through the Eco-clubs was, therefore, launched during 2001-2002. Under this programme, Eco-clubs were set up in 5,750 schools of each district of the Andhra Pradesh.

This programme is being implemented in five States: Andhra Pradesh, Delhi, Maharashtra, Kerala and Himachal Pradesh through the nodal agency appointed by the State/UT Government. Unless the steps are taken on a war footing, the future generation would suffer a lot. The Government of India provides financial assistance for establishment of Eco-clubs @ ₹ 2, 500 per Eco-club including, training of master trainers, teacher training and distribution of resource materials. Though this amount is meagre, even to spend this, the survey reveals that the school teachers are not coming forward. If the amount is raised and if the teacher is given some incentive per month or per annum, perhaps this Eco-club system among the schools may be effective. Not only this, an effective supervision by experts may be an added advantage to create environmental awareness. Apart from the above mentioned programmes, there are some more programmes introduced by the government to create environmental awareness not only in the schools but also in the society.

E. Other Awareness Programmes: The Ministry also sponsors various programs which do not fit into straitjacketed programs like NEAC, NGC, etc., and are aimed at creating environmental awareness among children. These include environment quiz (both written as well as televised), organization of activities for observation of special occasions such as Earth day, special programs for children, etc. These proposals which are received throughout the year from various NGOs and other agencies, are considered on merit as and when received and are supported. Organization of an annual vacation program on environmental resources for high school level students namely "Vacation Program on Natural Resources-Building a Broader Constituency of Support for Conservation" by ATREE, etc. is proposed for educating school children. There are also schemes for environmental education by the state governments in the country. We shall see the Andhra Pradesh government's activities below.

F. Environmental Education in Schools of Andhra Pradesh: It is being implemented since 2000, with the collaboration of Sarva Shiksha Abhiyan and Sir Ratan Tata Trust. The main objective of the Scheme is to develop a model for strengthening activity based environmental education in the schools. The thrust areas of the project are:

1. Environmental education for rural Schools-environmental education in schools of Andhra Pradesh; Bio-intensive gardens in schools.
2. Nature of education for urban Schools-environmental awareness campaign in summer coaching camps.
3. Organized by Greater Hyderabad Municipal Corporation-water and sanitation; participatory assessment of water and sanitation facilities in schools.

Environmental Education: Role of Teachers

So far as this Scheme is concerned, teachers can play a vital role in imparting environmental education to their students, which is possible only when the teachers themselves have the necessary level of awareness. The teachers should know the adverse impact of climate change and global warming at macro and micro levels. The urgency of imparting environmental education to the students through scientific experiments, which could be explained to peers, parents and communities during science exhibitions and community projects, are very much needed to understand the issues of environmental problems and the challenges to meet those problems in the near future. The challenges are many to be met.

Unless we train them at an infant stage, it is impossible, after a certain age, to meet them. The teachers have emphasized on the practice of conservation of water, energy and protection of trees at their personal level among the teachers as concerned citizens and also as role models to the students. The simple ways like switching off lights, electrical gadgets, closing water taps whenever not required, growing less water consuming trees, practicing traditional knowledge systems in farming in the school gardens etc. is essential to make them learn. The moral and professional responsibility of teachers in inculcating the ethical values such as care, compassion and responsibility towards mother Earth and her children among the students, is needed, especially these days.

The National Environmental Education Act of 1990 is an act of the Congress of the United States of America to promote environmental education. In this Act, Congress found that "threats to human health and environmental quality are increasingly complex, involving a wide range of conventional and toxic contaminants in the air and water and on the land" and that "there is growing evidence of international environmental problems, such as global warming, ocean pollution, and decline in species' diversity, and that these problems pose serious threats to human health and the environment on a global scale" and declared several other problems that need to be fixed or addressed by improving environmental education. Thus, the importance of education is recognized the world over.

Suggestions

1. Organizing seminars, workshops and training classes—

pupil participation is also recognized as an essential element in order to implement the strategies for sustainable agriculture, since the rural environment can only be protected with the active collaboration of the local population.

2. Now there is a good evidence of sunburn before the age of 15, which may increase the risk of skin cancer all around the world. Hence, there is an urgent need to target the children and teenagers advising them about the protection against sunburn.
3. Residential/non-residential tours to be conducted by the school management during vacations. While touring, teachers should teach them about farming, plantation programmes and how to use waste material for recycling, etc.
4. The school headmaster should organize an educational tour to any nearby forest to educate the children for the conservation of forests. This might enrich the students about the importance of forests.
5. With the help of any nearby ITI's, let the children learn the vocational courses during vacations like paper bags manufacturing, carpentry, painting, repairing etc. These activities may help the children to have a vision about the future.

   The above suggestions can improve the system of environmental education among school children. To make them perfect environmentalists, the following suggestions may also be useful at their budding age. To have a birthday to perform by a child, ask the children to bring recyclable items for the recycle bins; activities can include making artwork using recycled objects such as bottle caps, plastic lids and paper mache; or make bird feeders using recyclable plastic bottles.
6. Create a mini environmental education plot in your backyard, for your children, grandchildren, or neighbours' children. You might include a birdhouse, bird bath,

feeding station, rock piles, and logs; also plant flowers, herbs, or vegetables. Encourage the children to nurture the plot and to report on any changes and/or progress (the degree to which this can be done depends on the age of the children).

7. Explore the world of birds, butterflies, beetles or bats, or any other creature that is easy to observe in your area and that interests your child. All you'll need is binoculars and perhaps a guide book or other nature book. Your child may want to keep a log book of the different types of birds or number of bats observed, when they were seen, etc.
8. Water conservation project in schools may also help children understand the preservation of rain water. Now-a-days it is a scarce material.

To promote awareness and knowledge among children, many activities, suggestions mentioned above may be enough. If they are not sufficient, centres involving environment education can be set up with in school grounds. They can be taught about weather stations which show the temperature, rain fall, humidity, air pressure, wind blowing etc. Intense knowledge of geography, mathematics and science will make the students more active and creative.

This type of practical education is necessary to make the students active. The government is interested to inculcate such knowledge among the younger generations but the co-operation from the teachers appears to have been lacking. Every citizen of this country, especially, the teaching faculty must feel responsible to make the future citizens of this country environmentally effective and protective. We must realize the present status of environment and make adjustments with nature correctly for the sake of our own survival.

Degradation of the environment is going to affect each of us irrespective of the country, region or area. We should be concerned about our environment. Development without concern for the environment can only be a short-term development. Environmental issues emerge when you pursue the development. Challenges to meet them are also essential for development.

References

Government of India, Ministry of Human Resource Development (MHRD) (1986): National Policy on Education, New Delhi.

http://envfor.nic.in/divisions/ee/ee.html: Environment Education Division.

http://planningcommission.nic.in/reports/sereport/ser/separd/mchapter3.pdfChapter-3: Development schemes in details, HRD, Department, Environmental Orientation to School Education.

Jane Freak, (2007): "Evaluation of a sun awareness programme for school children", Vol. (103), Issue: 26, pp. 30-31

Kartikeya V. Sarabhai, Director, Centre for Environment Education, Ahmedabad, Paper; "Strategies in Environmental Education".

Pradeep Kumar Singhal and Pankaj Shrivastava, (2004): Sustainable Development, p. 127.

Public Law, 101-619, November 16, 199O: 10STAT: "National Environmental Education Act", pp. 3325-3340.

V. Swathi, (2007): The Hindu, Wednesday, July 4, "The National Green Corps (NGC) will be introducing 'Environmental Education Activity", Hyderabad.

10

Environmental Movements and Conflicts over Natural Resources

C. Basavaiah and I. Malyadri

Introduction

Natural resource economics deals with the supply, demand, and allocation of the Earth's natural resources. This subfield of economics is therefore interested in the primary sector of the economy which engages in resource extraction (that is, the extraction of raw materials). One main objective of natural resource economics is to better understand the role of natural resources in the economy in order to develop more sustainable methods of managing those resources to ensure their availability to future generations. Academic and policy interest has now moved beyond simply the optimal commercial exploitation of the standard trio of resources to encompass management for other objectives.

For example, natural resources more broadly defined have recreational, as well as commercial values. They may also contribute to overall social welfare levels, by their mere existence. The economics and policy area focuses on the human aspects of environmental problems. Traditional areas of environmental and natural resource economics, include welfare theory, pollution control, resource extraction, and non-market valuation, and also resource exhaustibility, sustainability, environmental management, and environmental policy.

The recent period in human history contrasts with all the earlier ones in its strikingly high rate of resource utilization. Ever expanding and intensifying industrial and agricultural production has generated increasing demands on the world's total stock and flow of resources. These demands are mostly generated from the industrially advanced countries of the North and the industrial enclaves in the underdeveloped countries of the South. Paradoxically, the increasing dependence of the industrialised societies on natural resources, through the rapid spread of energy and resource-intensive production technologies, has been accompanied by the spread of the myth that increased dependence on modern technologies implies a decreased dependence on nature and natural resources.

This myth is supported by the introduction of a long and indirect chain of resource utilisation which leaves invisible the real material resource demands of the industrial processes. Through this combination of resource intensity at the material level and resource indifference at the conceptual and political levels, conflicts over natural resources generated by the new pattern of resource utilisation are generally shrouded and overlooked. These conflicts become visible when resource and energy-intensive industrial technologies are challenged by communities whose survival depends on the conservation of resources threatened by destruction and overexploitation, or when the devastatingly destructive potential of some industrial technologies is demonstrated as in the Bhopal disaster.

For centuries, vital natural resources like land, water and forests had been controlled and used collectively by village communities thus ensuring a sustainable use of these renewable resources. The first radical change in resource control and the emergence of major conflicts over natural resources induced by non-local factors was associated with colonial domination of this part of the world. Colonial domination systematically transformed the common vital resources into commodities for generating profits and growth of revenues. The first industrial revolution was to a large extent supported by this transformation of commons into commodities which permitted European industries access to the resources of South Asia.

With the collapse of the international colonial structure and the establishment of sovereign countries in the region, this international conflict over natural resources was expected to be reduced and replaced by resource policies guided by comprehensive national interests. However, resource use policies continued along the colonial pattern and, in the recent past, a second drastic change in resource use has been initiated to meet the international requirements and the demands of the elites in the Third World, leading to yet another acute conflict among the diverse interests. The most seriously threatened interest, in this conflict, appears to be that of the politically weak and socially disorganised group whose resource requirements are minimal and whose survival is primarily dependent directly on the products of nature outside the market system. Recent changes in resource utilisation have almost wholly by-passed the survival needs of these groups. These changes are primarily guided by the requirements of the countries of the North and of the elites of the South.

This article analyses environmental conflicts in contemporary human society. In general it relates to societies all over the world, but in particular it addresses the most intense and emerging social contradictions in India related to conflicts over natural resources. Science and technology are central to these conflicts because while scientific knowledge has been used by contemporary societies to considerably enlarge man's access to natural resources, it has also allowed the utilisation natural resources at extremely high rates.

The contemporary period is characterised by the emergence of ecology movements in all parts of the world which are attempting to redesign the pattern and extent of natural resource utilisation to ensure social equality and ecological sustainability. Ecology movements emerging from conflicts over natural resources and the people's right to survival are spreading in regions like the Indian subcontinent where most natural resources are already being utilised to fulfil the basic survival needs of a large majority of people.

The introduction of resource and energy-intensive production technologies under such conditions leads to economic growth for a small minority while, at the same time, undermines the material basis for the survival of the large majority. In this way, ecology movements have questioned the validity of the dominant concepts and indicators of economic development. The ideology of economic development, which remained almost monolithic in the post World War II period, is thus faced with a major foundational challenge. In this chapter an attempt has been made to provide a systematic conceptual framework for analysing the processes and structures of modern economic development from an ecological-perspective. It attempts to analyse the relationship between economic development and conflicts over natural resources to trace the roots of ecological movements. Further, in the light of the ecological perspective, it examines the fundamental assumptions and categories of modern development economics that are used to determine the objectives of economic development as well as the criteria for the choice of technologies that are used to achieve these objectives.

Development and Environmental Conflicts in India

A characteristic of Indian civilization has been its sensitivity to natural ecosystems. Vital renewable natural resources like vegetation, soil and water were managed and utilised according to well defined social norms that respected the known ecological processes. The indigenous modes of natural resources utilisation were sensitive to the limits to which these resources could be used It is said that the codes of visiting important pilgrim centres Badrinath in the sensitive Himalayan ecosystem, included a maximum stay of one night so that the temple area would not put excess pressure on the local natural resources base. In the pre-colonial indigenous economic processes, the levels of utilisation of natural resources were not significant enough to result in drastic environmental problems.

There were useful social norms for environmentally safe resource utilisation and people protested against the destructive use of resources even by kings. A major change in the utilisation of natural resources of India was introduced by the British who linked the resources of this country with the direct and large non-local demands of Western Europe. Natural resource utilisation by the East India Company, and later by the colonial rulers, replaced the indigenous organizations for the utilisation of natural resources, like water, forest and minerals that were mainly managed as commons.

Three Economies of Natural Resources

A new and holistic relationship between economics and ecology has to depend on a holistic understanding of the natural resource process and utilizations associated with human societies and the natural ecosystems. The dominant ideology of development which guides development activities almost exclusively has been classically concerned only with the use of natural resources for commodity production and capital accumulation. It ignores the resource processes that have been regenerating natural resources outside the realm of human existence.

It also ignores the vast resource requirements of the large number of people whose needs are not being satisfied through the market mechanisms. The ignorance or neglect of these two vital economies of natural resources, the economy of natural processes and the survival economy, explains why ecological destruction and threat to human survival have remained hidden negative externalities of the development process. To make good for this shortcoming it is necessary to comprehend the place of natural resources in all the three economies.

Natural Resources in the Market Economy

The incompetence of modern economics in dealing with natural resources in their ecological totality has been voiced by many. The most penetrating description, however, comes from Georgescu Roogen who wrote: The no deposit no return analogy benefits the businessman's view of economic life. For, if one looks only at money, all one can see is that money just passes from one hand to another: except by regrettable accident it never gets out of the economic process. Perhaps the absence of any difficulty in securing raw materials by those countries where modern economics grew and flourished was yet another reason for economists to remain blind to this crucial economic factor.

Economy of Natural Ecological Processes

Natural resources are produced and reproduced through a complex network of ecological processes. Production is an integral part of this economy of natural ecological processes but the concepts of production and productivity in the context of development economics have been exclusively identified with the industrial production system for the market economy. Organic productivity in forestry or agriculture has also been viewed narrowly through the production of marketable products of the total productive process. This has resulted in vast areas of resource productivity, like the production of humus by forests, or regeneration of water resources, natural evolution of genetic products, erosional production of soil fertility from parent rocks, remaining beyond the scope of economics.

Many of these productive processes are dependent on a number of ecological processes. These processes are not known fully even within the natural science disciplines and economists have to make tremendous efforts to internalize them. Paradoxically, through the resource ignorant intervention of economic development at its present scale, the whole natural resource system of our planet is under threat of a serious loss of productivity in the economy of natural processes. At present ecology movements are the sole voice to stress the economic value of these natural processes. The market-oriented development process can destroy the economy of natural processes by over exploitation of resources or by the destruction of ecological processes that are not comprehended by economic development.

And these impacts are not necessarily manifested within the period of the development projects. The positive contribution of economic growth from such development may prove totally inadequate to balance the invisible or delayed negative externalities stemming from damage to the economy of natural ecological processes. In the larger context, economic growth can thus, itself become the source of underdevelopment. The ecological destruction associated with uncontrolled exploitation of natural resources for commercial gains is a symptom of the conflict between the ways of generating material wealth in the economies of market and the natural processes. In the words of Commoner: "Human beings have broken out of the circle of life driven not by biological needs, but the social organisation which they have devised to 'conquer' nature: means of gaining wealth which conflict with those which govern nature".

In conclusion, modern economics and the concept of development cover a miniscule portion in the history of economic production by human beings. The survival economy has given human societies the material basis of survival by deriving livelihoods directly from nature through self-provisioning mechanisms. In most Third World countries, large number of people are deriving their sustenance in the survival economy in ways that remain invisible to market oriented development. Within the context of a limited resource base the destruction of the survival economy takes place through the diversion of natural resources from directly sustaining human existence to generating growth in the market economy. Sustenance and basic needs satisfaction is the organising principle for natural resource use in the survival economy whereas profits and capital accumulation are the organising principles for the exploitation of resources for the market. Human survival in India even today is largely dependent on the direct utilisation of common natural resources.

11

Judicial Activism for Environment

J.V. Siva Kumar

Right from the mother's womb, one needs unpolluted air to breathe, uncontaminated water to drink, nutritious food to eat and hygienic conditions to live in. These elements are *sine qua non* for sound development of human personality.

Right to live is a fundamental right under Article 21 of the Constitution and it includes the right of enjoyment of pollution free water and air for full enjoyment of life. If anything endangers or impairs that quality of life in derogation of loss, a citizen has the right to have recourse to Article 32 of the Constitution for removing the pollution of water or air, which may be detrimental to the quality of life.

The Constitution (Forty-second Amendment) Act, 1976 introduced Articles 48 in Part IV which provides that "the State shall endeavour to protect and improve the environment and safeguard the forests and wildlife of the country". This provision though not enforceable in a court, directs the State to enact legislation and frame policies towards attaining these goals. The Forty-second Amendment also introduced Article 51A in Part IVA which casts a fundamental duty on the citizen "to protect and improve the material environment including forests, lakes, rivers and wildlife, and to have compassion for living creatures". Thus, the state now is under a moral duty to take measures to prevent ecological imbalances resulting from modern industrialization. The Constitution has also cast a duty on the citizen to take steps for maintaining ecological balance.

The turmoil and confusion following the Bhopal Gas leakage has brought about into light the government's total lack of concern over the environment. It is only after the

Bhopal tragedy and the ensuing publicity that the public has become aware of environmental 'pollution' and its devastating effect and the fact that it is every citizen's right to live in a pollution free and healthy environment.

The tide of judicial activism in environmental litigation in India symbolizes the anxiety of Courts in finding out appropriate remedies for environmental maladies. At the global level, the right to live is now recognized as a fundamental right to an environment adequate for health and well-being of human beings. In the context of such developments in the international scenario, the judicial decisions in India are particularly worth examining.

"Protection of environment is not only the duty of the citizens but also the obligation of the State and all other State organs including the Courts. To that extent, environmental law has succeeded in unshackling men right to life and personal liberty from the clutches of common law theory of individual ownership".

The powers of the Magistrate under the Code of Criminal Procedure to move against public nuisance has been developed by the judiciary as a significant instrument of environmental protection at the grass root level. Different are the situations in the which the power can be used, from removing nuisance created by the oven and chimney of a bakery and closing a factory and its boiler causing noise and air pollution in a residential locality to even directing a self-governing local body to construct drainage and public lavatories. Municipal Council, Ratlam vs. Vardhichand, is milestone in this path rejecting the local council's plea of financial inability for providing basic amenities to the people. The Supreme Court tried to find in the law of public nuisance a social justice component of the rule of law and consider it as viable instrument of environmental protection in the third world countries.

According to the Court, the law operates against statutory bodies and other regardless of the cash in their coffers because human rights are to be respected by the State regardless of budgetary provision. The message of the decision is that the duty to provide the basic amenities to the people cannot be evaded even if the local self-governing bodies can justify their existence.

Environmental Protection through Public Interest Litigation

Environmental policy making, like other controversial areas of public, is a never-ending process and the courts will always be intimately involved in this process. The increase in environmental awareness since 1980s has triggered a spurt in the environmental cases reaching the courts. It is emphatically, the province and duty of the judiciary, to say what the law is. Most of these actions in environmental cases are brought under Article 32 and 226 of the Indian Constitution, since the litigants and the lawyers prefer the direct access to the nation's highest courts that the writ provides. The environmental petitions are generally based on the plea of violation of fundamental rights. It was not until 1978 that the Supreme Court breathed substantive life into Article 21 by subjecting state action interfering with life or liberty to a test of reasonableness; requiring not only that the procedures be authorized by law, but that they are "right, just and fair". This transformation paved the way for a substantive reinterpretation of constitutional and legal guarantees and positive judicial intervention.

Right to a Wholesome Environment

Encouraged thus by an atmosphere of freedom and articulation, the Supreme Court rejected the "bureaucratic tradition" of mechanical and rule-bound adjudication and entered one of its most creative periods. Most significantly, the Court fortified and expanded Article 21 of the Indian Constitution. In the process, the right to a wholesome environment (although not explicitly mentioned in Part III) was drawn within the expanding boundaries of the fundamental right to life and personal liberty, guaranteed in Article 21.

In Dehradun Quarrying Case, for the first time, the Supreme Court held that the fundamental right to a wholesome environment is a part of the fundamental right to life in Article 21 of the Constitution. In July, 1983, the representative of the Rural Litigation and Entitlement Kendra, Dehradun wrote to the Supreme Court alleging that illegal limestone mining in the Mussorie-Dehradun Region was devastating the fragile ecosystems in the area. On 14th July, 1983, the Court directed the registry to treat the letter as a writ petition under Article 32 of the Constitution with notice to the Government of Uttar Pradesh and the Collector of Dehradun. Over the years, the litigation grew complex.

The Court delivered its final judgment in this case in August, 1988 after hearing lengthy argument from central and state governments, government agencies and mine lessees; appointed several expert committees; and passed several comprehensive; interim orders. None of these orders, however articulate the fundamental rights infringed. The exercise of jurisdiction under Article 32 presupposes the violation of a fundamental right. Therefore, it was necessary to reasonably hold that enjoyment of right to life under Article 21 of the Constitution embraces the protection and preservation of a wholesome environment without which life cannot be enjoyed. This view also supported by Justice Singh's concluding observations justifying the closure of pollution tanneries in the Ganga Pollution case, "we are conscious that closure of tanneries may bring unemployment, loss of revenue, but life, health and ecology have greater importance to the people".

In M.C. Mehta (2) vs. Union of India, the Supreme Court has held that under Article 51-A(g), it is the duty of the Government to introduce compulsory teaching of lessons at least for one hour in a week on protection and improvement of natural environment in all the educational institutions of the country. It directed the Central Government to get text books written on that subject and distribute them to the educational institutions free of cost in order to create awareness amongst the people. Regards the consciousness of cleanliness of environment, it suggested the desirability of organizing 'keep the city clean week', 'keep the town clean', 'keep the village clean week' in every city, town and village throughout India at least once in a year.

In the case of M.C. Mehta vs. Union of India, 1997, 2 SC.C.353, popularly known as Taj Mahal case, the judgment of the court was based on the "principle of sustainable development and the court applied the precautionary people". In this case, public interest litigation was filed alleging that due to environment pollution there is a degradation of Taj Mahal, a monument of international reputation. According to the opinion of the expert committees, the use of coke/coal by industries situated within the Taj Trapezium Zone (TTZ) were emitting pollution and causing damage to the Taj Mahal.

In M.C. Mehta v. Komal Nath, 2002, 3 SC.653, keeping in view various judgments given by the court in Public Interest Litigations (PIL) the court gave the punishment in this case for violation of ecological imbalance. The court further ordered for quantification of exemplary damages already awarded payable on the principle of "polluter pays".

In M.C. Mehta vs. Union of India (UOI) and others, on behalf of Monitoring Committee, the Supreme Court directed CBI to lodge an FIR and make further investigations in accordance with law. The court in its earlier orders regarding pollution of atmosphere in the vicinity of Taj Mahal had appointed a committee to report on the progress of the action being taken by Agra Mission Management Board. CBI was directed to conduct enquiry into the matter of development of heritage corridor by NPCC. It was also directed to interrogate the persons involved and to verify their accounts as it was alleged that the concerned officials and the Chief Minister had released ₹ 17 crore without proper sanction. A detailed report was submitted to Supreme Court on this issue. Direction was given to the Central Government as well as to the state government to hold departmental enquiry against the concerned secretaries as well as the managing director of NPCC who had undertaken the project.

The Supreme Court in 2005 dealing with five noise pollution cases in an exhaustive manner, explained that any noise which has the effect of materially interfering with ordinary comforts of life, judged by the standards of a reasonable man, is a nuisance, and a nuisance created by noise becomes actionable depending on its degree, surrounding circumstances the place and time. The people in general have the right to freedom from noise pollution. This is fundamental right protected by Article 21, and noise pollution beyond permissible limits is an inroad into that right. Noise constitutes a real and present danger to people's health.

From the foregoing, it is clear that the judicial review of legislative measures, governmental programmes and administrative decisions can be an effective instrument of enforcement matters. The Supreme Court has made significant contribution in giving a fillip to the citizen's right to a hygienic environment, but the exercise of their discretionary powers in environmental matters is yet to take a concrete form. The courts have time and again faced difficulties in respect of investigating machinery required for the citizens' suits in environmental matters. As a result, the difficulty has been felt about the availability of authentic data. To overcome this, the courts have resorted to appointing the distinguished persons as experts or commissioners to investigate and report to it. This ad-hoc practice, however, needs to be put on a permanent footing, preferably as a special cell, to meet the growing number of environmental litigations expected in the coming years. In this context, it is suggested here that the environment courts on a regional basis, with one professional judge and two experts be drawn up and ecological sciences research group, should be set up.

The right to environment is not merely a moral concern. It is a comprehensive right like any other basic right at both national and international levels. The Indian judiciary has interpreted the various constitutional and legal provisions relating to environment in an appropriate direction by promoting ecological balance and sustainable development.

References

Jamie Cassels, "Judicial Activism and Public Interest Litigation: Attempting the Impossible?" The American Journal of Comparative Law: 4/501-4 (1989).

12

Global Environmental Issues in the 21st Century

Dhulasi Birundha Varadarajan

Introduction

The key environmental issues in the 21st century may result from unforeseen events and scientific discoveries, sudden, unexpected transformations of old issues, and well known issues that do not receive enough policy attention. A survey of emerging issues carried out among scientists for Global Environment Outlook (GEO), 2000 cited pollution and scarcity of water resources (57 percent) as major issues. Then came deforestation/problem arising from poor governance at national and international levels (27 percent).

Environmental problems are climbing even higher on the international political agenda at times; preoccupying international diplomats almost as much as arms control negotiations did during the Cold War. Industrial countries are increasingly arguing, with the European Union (EU) and the United States at odds on issues from global climate change to genetically modified organisms (GMOs). Environmental issues have also become acrimonious in North-South relations, with rich and poor countries divided over how to apportion responsibility for reversing the planet's ecological decline (Hilary French, 2000).

The Millennium Ecosystem assessment conducted by 1360 experts from 95 countries concluded that about 60 percent of the ecosystem services that support life on Earth are being degraded or used unsustainably. The current absorption capacity of carbon by oceans and forests is about 3-3.5 billion tons per year. Today, 7 billion tons are added annually.

Glaciers are receding world wide and Arctic ice is melting faster than expected. Atmospheric CO_2 which for 40,00,000 years fluctuated between 180 and 280 ppm has risen to 380 ppm. Plants are blooming at an average of 5.2 days around the world than they did 10 years age. The Intergovernmental Panel on Climate Change estimates a 1.4-5.8^{0}C (2.5-10.4^{0}F) warming by Century's end, which could raise sea levels by 88 centimetres (35 inches) changing human coastal settlements. FAO estimates that agricultural land lost due to climate change could cost as much as US$ 56 billion per year. An estimated 90 percent of the total weight of the ocean's large predators, tuna, sword fish and sharks–has disappeared in recent years. An estimated 12 percent of bird species, 25 percent of mammals, and more than 30 percent of all amphibians are threatened with extinction within the next century. Climate change may threaten more than 1 million species with extinction by 2050. Half the world's forests and 25 percent of the coral reefs are gone (Jerome C. Glenn and Theodore J. Gordon, 2005).

Climate Change

The climate change policies can be broadly divided into two categories: GHG mitigation strategies and adaptation strategies. The GHG mitigation strategies dominate the current global negotiations on climate change and the Kyoto Protocol signed at the Third Conference of Parties in 1997 has for the first time placed quantitative restrictions on the GHG emissions from the developed countries. However, the implementation of Protocol continues to be surrounded by, much uncertainty and Prins and Rayner (2007) in their recent influential article have argued that if the required fundamental changes are needed in the world climate regime, the world should look beyond Kyoto Protocol.

The global average temperature has increased by about 0.7°C (13°F) during the 20th century. The Intergovernmental Panel on Climate Change (IPCC) concluded in 2001 that humans are already having a discernable impact on the global climate—most of the observed warming over the last 50 years is likely to have been due to the increase in green house gas concentrations. In 2007 they reaffirmed and challenged this conclusion. Current emissions trends will lead to a doubling of green house gas concentration over pre-industrial levels by around 2050. The IPCC projects a global average temperature increase of 1 to 6 degrees centigrade, or 2 to 10 degrees Fahrenheit by 2100. This would have significant impact on climate throughout the world (Cline 1992, Frank Hauser 1995, IPCC 2001, 2007).

Global emission of carbon dioxide from combustions of fossil fuels rose dramatically during the 20th century. The use of petroleum is currently responsible for about 42 percent of global carbon emissions, while coal is the source of another 36 percent. The United States is presently the world's largest emitter of CO_2. The global carbon dioxide emissions are projected to increase by approximately 59 percent between 2004 and 2030 and in China by about 140 percent. Although carbon emissions are projected to grow fastest in the developing nations, per capita emissions in 2020 would still be much higher by about six terms in the industrialized countries.

The Earth has warmed significantly over the last 100 years. The global average temperature has risen by about 0.7°C or about 1.3°F. Global temperature since 2000 has been particularly warm—six of the seven warmest years on record have occurred since 2000. There is also evidence that the rate of warming currently about 0.13°C per decade is increasing. The IPCC estimates that during the 21st century, global average temperature will rise between 1.1°C (2°F) and 6.4°C (11°F) with the range more likely to be between 1.8°C (3°F) and 4°C (7°F).

The magnitude of actual warming and other effects will depend upon the level at which atmospheric concentrations of CO_2 and other green house gases are ultimately stabilized. The current atmospheric CO_2 concentration is around 380 ppm.

When we consider the contribution of other green house gases, the overall effect is equivalent to a concentration of 430 ppm of CO_2. Stabilizing green house gas concentrations at 450 ppm Co_2 would be 90 percent likely to eventually result in a temperature increase between 1°.0 and 3.8°C (Jonathan M. Harris and Brain Roach).

Water Crisis

Global freshwater consumption rose six fold between 1900 and 2000, more than twice the rate of population growth, and the rate of increase of consumption is still accelerating. Demand for water resources is increasing both because of population growth (particularly in developing countries) and because of rising demand per person due to such causes as irrigation development, industrialization and increasing use by individuals, as incomes rise. A potential crisis is looming where available resources can no longer meet the needs.

Water stress is said to exist when annual per capita availability at national level is below 1,600m^3 per year, for all uses, including the cultivation of food. A level below 1,000m^3 is regarded as absolute security. Currently about 30 countries, including Israel, Jordan and South Africa, are considered water stressed and 20 face absolute water scarcities' (Allan, 200).

In 2000, the International Water Management Institute (IWMI) concluded that by 2025, 33 percent of the world's population, or two billion people, will be living in countries or regions with absolute water scarcity, using the UN medium population growth projections. Most countries in the Middle East and North Africa will have absolute scarcity and would be joined by Pakistan, South Africa and large parts of India and China (Mehta, 2000).

By far the largest consumer of water is the cultivation of food. IWMI predicts that the world will have to provide an additional 22 percent of primary water to meet the future food needs by 2025, nearly three quarters of which will be for irrigation. Even given this increase, the International Food

Policy Research Institute (IFPRI) predict a virtual doubling in food imports in the next 20 years in order to fill the local food production-demand gap.

Currently, some 1.2 bn people lack access to water and 3.3 bn have no effective sanitation, which causes or contributes to the death of more than 3mn people each year from water-related diseases. Population growth ensures that the demand for water and sanitation continues to rise, to the extent that by 2025, an estimated additional 3.1 bn people will need access to water. The tendency to give higher priority to urban service delivery has already contributed to a disparity in service levels between rural and urban areas—in sub-Saharan Africa, 77 percent of the urban population is served against 39 percent rural and the global averages are respectively, 90 percent for urban and 62 percent for rural areas. The major initial impetus was the UN Water Conference held in Mar del Plata in March 1977, at which parties agreed to declare the 1980s as the International Decade of Water supply and Sanitation. The 'Water Decade' as it came to be known, set the ambitious target of 'water for all' by 1990, and aimed to achieve this through emphasizing government action and community initiative. By 1990, though progress had been made, billions of people remained without such access.

'Bringing in' the private sector for water supply and water resource development, from the global to local level is a major current narrative. Six major transnational companies presently dominate the water utility market and by 2010 are poised to increase their overall share in markets worth some US$ 20 bn in Africa, Asia and Latin America, by which time the private sector will account for between 20 percent and 60 percent of all water and sanitation services supplied in these countries.

Land Degradation

Land degradation is broadly defined as "...any form of deterioration of the natural potential of land that affects ecosystem's integrity either in terms of reducing its sustainable ecological productivity or in terms of its native biological richness and maintenance of resilience (GEF 1999)". It is a worldwide phenomenon substantially affecting productivity in over 80 countries on all continents, except Antarctica. Land degradation is especially serious in Africa where 36 countries face dry land degradation or desertification. Land degradation damages soil structure and leads to the loss of soil nutrients through processes such as water or wind erosion, water logging and salinization, and soil compaction. The main causes of land degradation are inappropriate land use, mainly unsustainable agricultural practices, overgrazing, and deforestation.

Land degradation adversely affects the ecological integrity and productivity by about 2 billion ha or 23 percent of landscape under human use. Agricultural land in both dry land and forest areas have been most severely affected by land degradation. They cover about one-fourth of the world's total land area and account for 95 percent of all animal and plant protection and 99 percent of calories consumed by people. About two-thirds of agricultural lands have been degraded to some extent during the last 50 years (WRI, 2000). This impact has already put at risk the livelihood and economic well being, and the nutritional status of more than 1 bn people in the developing countries.

The World Summit on Sustainable Development (WSSD, 2002) reaffirmed land degradation as one of the major global environment and sustainable development challenges of the 21st century, calling for action to "...address causes of desertification and land degradation in order to restore land, and to address poverty resulting from land degradation". The Summit also emphasized that "sustainable forest management of both natural and planted forest and for timber and non-timber products is essential in achieving sustainable development and is critical factor in attaining the Millennium Development Goals of reducing by half the proportion of people in poverty by 2015 and ensuring environmental sustainability".

The three post Rio (UNED, 1992) global environment conventions that are most relevant to land degradation prevention and control are: the UNCCD, Convention of Biological Diversity (CBD), and the United Nations Framework Convention on Climate Change (UNCCC). Consensus on land degradation as a global environment and sustainable development issue, as well as the need for coordinated international action to address it, led to the adoption of the UNCCD in June 1994. The objective of the UNCCD was to "...combat desertification and mitigate the effects of countries experiencing serious drought and/or desertification, particularly in Africa, through effective actions at all levels, supported by international cooperation and partnership arrangements, in the framework of an integrated approach which is consistent with Agenda 21, with a view to contributing to the achievement of sustainable development in the affected areas". The CBD recognizes the importance of addressing land degradation. For example, the program priorities of the convention highlights the role that land degradation and deforestation prevention and control can lay in the conservation and sustainable use of biodiversity.

Forest

To mention the latest state of forests (Report of the Forest Survey of India), the actual forest cover of India is 19.27 percent of the geographical area, corresponding to 63.3 million ha. Only 38 million ha of forests are well stocked (crown density above 40 percent). This resource has to meet the demand of a population of 950 million people and around 450 million cattle. As such, country has to meet the needs of 16 percent of the world's population from 1 percent of the world forest resources. The same forest has also to cater for the 19 percent of the cattle population.

The forests of the country are therefore, under tremendous pressure. Forest fires are a major cause of degradation of India's forests. While statistical data on fire loss are weak, it is estimated that the proportion of forest areas prone to forest fires annually ranges from 33 percent in some states to over 90 percent in others. About 90 percent of the forest fires in India are created by humans. The normal fire season in India is from the month of February to mid of June. India witnessed the most severe forest fires in the recent times during the summer of 1995 in the hills of Utter Pradesh and Himachal Pradesh. The fires were very severe and attracted the attention of whole nation. An area of 677,770 ha was affected by fires.

The Forest Survey of India data on forest fires attribute around 50 percent of the forest areas as fire prone. This does not mean that country's 50 percent area is affected by fires annually. Very heavy, heavy and frequent forest fire damages are noticed only over 0.8 percent, 0.14 percent and 5.16 percent of the forest areas respectively. Thus, only 6.17 percent of the forests are prone to severe fire damage. In absolute terms, out of the 63 million ha of forests an area of around 3.73 million ha can be presumed to be affected by fires annually.

An analysis of macro economic impacts of natural disasters carried for the United Nations World Wide Disaster Report, 2003 showed that between 1991 and 2000, over 6,65,000 people were affected in 2,557 natural disasters, of which 90 percent were water related events and with 97 percent of the victims from developing countries.

In 2004, one of the largest earthquakes in recorded history (measuring 9 on the Richter scale), struck just off Sumatra, Indonesia, in a line running under the sea. The rupture caused massive waves, or tsunamis, that hurtled away from the epicentre, reaching shores as far as Africa. At least 230,000 people were killed and the livelihoods of millions were destroyed in over 10 countries. This has been one of the biggest natural disasters in recent human history.

Problem of Alien Species

The globalization of trade and the power of the internet

offers new challenges to those seeking to control the spread of IAS, as sales of seeds and other organisms over the internet pose serious new risks to the bio security of all nations. Controls on both import and export of species are required as part of a more responsible attitude of governments toward the potential spreading of invasive species around the world. While receiving countries, must ensure that they are able to control the imported species, few countries yet have effective controls in place. Because global trade has such a profound influence by moving species around the world, it is particularly important to ensure that concerns about IAS are built into relevant trade negotiations. Initial efforts are being made in this regard. For example, the Bio Safety Protocol under the CBD is part of the global trade regime; it is to be mutually supportive of any agreements under the World Trade Organization (WTO). The protocol is based on the principle that potentially dangerous activities can be restricted or prohibited even before they can be scientifically proven to cause serious damage, whereas decisions under trade law typically require "sufficient scientific evidence" to lead to such restrictions. In any case, IAS is so important that they should form part of the WTO agenda.

Changes in climate may also produce more conducive conditions for the establishment and spread of invasive species, as well as change the suitability of local climates for native species and the nature of interactions among native communities. The greatest impact of climate change on invasive species may arise from changes in the frequency and intensity of extreme climatic events that disturb ecosystems, making them vulnerable to invasions, thus providing exceptional opportunities for dispersal and growth of invasive species.

Agricultural and Invasive Alien Species

Unintentional introduction of pests and diseases of agriculture as contaminants in crops and animals has led to particularly severe problems, because alien species thrive in new ecosystems where their hosts are abundant and their own natural controlling factors may be absent. In the USA, approximately 15 percent of IAS cause severe damage; documented losses caused by 43 invasive alien insects for the period between 1906 to 1991 has been estimated at over US$ 92.5 billion (U.S Congress, Office of Technology Assessment, 1993). Recent increase in South-South and North-South horticultural trade has greatly increased the movement of IAS into forests, including pests such as whiteflies, leaf miners and trips.

Species Extinction

Human activity was cited as the cause of 816 plant and animal species having vanished in the past 500 years. However, the conservation group cautions that our knowledge of how many species exist—or used to exist—is still partial. The 5,611 threatened plants currently listed as threatened may represent only a small fraction of the number of species truly under attack since the group estimates that only 4 percent of all known plant species have been fully evaluated—and many more have not yet been discovered. Green turtle females lay upward of 100 eggs yearly on Caribbean beaches in Central America–but today just one of those eggs will grow into an adult turtle. The species is just one the 11,046 plants and animals that risk disappearing forever, according to the most comprehensive analysis of global conservation ever undertaken, the World Conservation Union's 2000 Red List of threatened species. The Report released, examined some 18,000 species and sub-species around the globe. Earth has estimated 14 million species-and only 1.75 million have been documented.

Conservationists estimate that the current extinction rate is 1,000 to 10,000 times higher than it should be under natural conditions. The 2001 IUCN Red List of Threatened Species is released once every four years by IUCN–the World

Conservation Union. The Red List is considered the most authoritative and comprehensive status assessment of global biodiversity. The fact that the number of critically endangered species has increased–mammals from 169 to 180; birds from 168 to 182 was a jolting surprise, even to those already familiar with today's increasing threats to biodiversity. These findings should be taken very seriously by the global community," says Maritta von Bieberstein Koch-Weser, IUCN's director general. The number of critically endangered primates increased from 13 to 19 endangered primates number to 46 today, up from 29 four years ago. Russell Mittermeier, President of Conservation International and chair of IUCN's Primate Specialist Group says, "The Red List is a solid documentation of the global extinction crisis, and it reveals just the tip of the iceberg. Many wonderful creatures will be lost in the first few decades of the 21st century unless we greatly increase levels of support, involvement and commitment to conservation".

Coral Bleaching

In the mid-1980s, coral bleaching was more severe than ever before and occurred in at least 60 countries (ISRS, 1999). Nearly 60 percent of the Earth's coral reefs are threatened by human activity, ranging from coastal development and over fishing to inland pollution leaving much of the world's marine biodiversity at risk.

A map-based indicator of threats to the world's coral reefs, reports that even though reefs provide billions of people and hundreds of countries with food, tourism revenue, coastal protection and new medications for increasingly drug resistant diseases—a set of goods and services worth about US$ 375 billion each year—they are among the least monitored and protected natural habitats in the world. Coral reefs of Southeast Asia, the most species-rich on Earth, are the most threatened of any region. More than 80 percent are at risk, primarily from coastal development and fishing-related pressures. The detailed, map-based analysis included in Reefs at Risk Report, serves as an indicator of the threats to coral reef ecosystems, not an actual measure of degradation. Until now, the only information of the status of coral reefs worldwide was a 1993 estimate, based on guesswork by a number of scientists, and anecdotal evidence, which indicated that 10 percent of the world's reefs were dead, and 30 percent were likely to die within 10 to 20 years. Using a Geographic Information System (GIS), more than 14 types of global maps, information on sites known to be degraded, and input from top coral reef scientists from around the world to model areas where reef damage is predicated to occur, Reefs at Risk documents that most coral reefs are seriously threatened by human activity.

Strategies Suggested

1. Design policies that favour alternative energy use, with differentiated responsibilities for developed and developing counties in relation to the equitable use of the global atmosphere.
2. Develop international strategies for de-carbonization.
3. Promote sustainable development as the central theme in policies relating to agriculture, trade, investment, research and development, infrastructure and finance by stressing the high economic and social value of environmental goods and services, and the high costs of poor environmental management.
4. Conduct more research on the socio-economic causes of environmental deterioration and the interlinkages within and among environmental and sustainability issues in order to define the priority issues and suggest ways of addressing them.
5. Establish a multi-agency, multi-stakeholder task force to develop proposals for strengthening global coordination and governance structure to protect the global commons.
6. Making full use of economic instruments that treat land water as scarce economic resources that are part of the

Earth's natural capital.

7. Coordinating the management of land and water resources as closely as possible.
8. Establishing secure land and water property rights where these do not exist.
9. Strengthen national cross-cutting institutions while maintaining strong environment agencies able to implement environmental policies and enforcing environmental laws and assess the overall state of the environment.
10. Increase support for international environmental organizations to enable them to improve their advisory, coordinating, and mediating policies.

If the world is to effectively address the pressing environmental problems on the global agenda, stronger international governance will be needed. Stronger international governance will evolve both from better utilization of existing mechanisms, and overtime through the development of new ones. The synergy between economic growth and technological innovation has been the most significant engine of change for the last 200 years, but unless we improve our economic, environmental and social behaviour, the next 200 years could be different. Thus, the achievement of sustainable development is a global endeavour which requires a massive reorganization of the world's economic policies, institutions and technology, and could be possible only if there is co-ordination of all sectors and all countries. It requires a universal effort and cannot be achieved if just one country follows sustainable path; for the gains to the world environment due to one country's sustainable development will be cancelled out by the environmentally destructive practices of others: on spaceship earth all have to sink or swim together.

References

Alan, J. A (2001) '*The Middle East Water Question–Hydropolitics and the Global Economy,* I.B. Tauris: London.

Cline, William R. (1992), The Economics of Global Warming, Washington D.C, Institute for International Economics.

Frank Hauser, Samuel (1995), Valuing Climate Change: the Economics of the Green House, London: Earth Scans Publications.

GEF (1999), Report of the STAP Expert Group Workshop on Land Degradation (GEF/C.14/Inf.15).

Hilary French, (2000), Coping with Ecological Globalisation-State of the World, 2000, Chapter 10, pp. 185-201.

ISRS (1999), Statement on Global Coral Bleaching in 1997-98, International Society for Reef Studies, Florida Institute of Oceanography, Florida, United States.

Intergovernmental Panel on Climate Change (2001) IPCC, Climate Change 2001, Vol. 1, The Scientific Basis, Cambridge, U.K., Cambridge Press, 2001.

Intergovernmental Panel on Climate Change (2007) IPCC, Climate Change 2007, Vol. 1, The Physical Science Basis, Available at http:// www.ipcc.ch.

Jonathan M. Harris and Brain Roach, (2007), The economics of Climate Change, Global Development and Environment Institute, Tuffs University M.A.

Mehta, L. (2000), Water for the 21st Century–Challenges and Misconceptions, IDS *Working Paper* 111, Institute of Development Studies: Briton.

Prins, G. and S. Rayner (2007), Time to Ditch Kyoto, Nature, 449, pp. 973-975.

United Nations Conference in Environment (1992), Rio de Janeiro, Brazil, June (1992).

World Summit on Sustainable Development (2002), Plan of Implementation, September 2002.

Appendix

Appendix

National Conservation Strategy and Policy Statement on Environment and Development

(Ministry of Environment and Forests, Government of India), June 1992

We are in the last decade of an extra-ordinarily eventful twentieth century. The world has seen spectacular political, social, cultural, economic and scientific progress during this century. But this progress has been monopolized by the chosen few at the unbelievably and indescribably large cost of the majority of mankind. The most disconcerting manifestation of this lop-sided progress has been our planet's ravaged ecology.

A good environmental sense has been one of the fundamental features of India's ancient philosophy. However, during the last few decades global circumstances have forced our country into a situation where it is becoming increasingly difficult to practice a life style that does not push this planet towards doom. During the last ten years, there has been a gratifying resurgence of this good environmental sense in this country. The most important aspect of this growing environmental consciousness in this country is its permeation at the establishment as also the people's level.

It is imperative that environmental consciousness becomes a pre-occupation with our people as no amount of government intervention can reverse ecological collapse. I see clear signs of that happening in India. Against this backdrop, we now have a system of environmental checks and balances fully in place. There is enough institutional, legislative and political strength to combine with a responsive citizenry to produce a practicable environmental culture. In Constitutional terms too,

India has enough guarantees to protect its ecological systems.

Since the inception of this Ministry, we have evolved enough to be able to chart out a life, which is happy without compromising the environment. There is a sizeable number of people who can rein in an indiscriminate establishment. In fact, we are now working towards a unique compatibility between the Development and the Environment.

We have our great past to draw from to create an equally great future. I see this environmentally degraded present only as an aberration for an enlightened civilization.

What you will read in the following pages are some of the specific means through which we propose to attain the goals of an environmentally-wise society.

1.0 Preamble

1.1 The survival and well-being of a nation depend on sustainable development. It is a process of social and economic betterment that satisfies the needs and values of all interest groups without foreclosing future options. To this end, we must ensure that the demand on the environment from which we derive our sustenance, does not exceed its carrying capacity for the present as well as future generations.

1.2 In the past, we had a great tradition of environmental conservation which taught us to respect nature and to take cognizance of the fact that all forms of life— human, animal and plant—are closely interlined and that disturbance in one gives rise to an imbalance in other's. Even in modem times, as is evident in our constitutional provisions and environmental legislation and planning objectives, conscious efforts have been made for maintaining environmental security along with developmental advances. The Indian Constitution has laid a new important trail in the Section on Directive Principles of State Policy by assigning the duties for the State and all citizens through Article 48 A and Article 51 A (g) which state that the "State shall endeavour to protect and improve the environment and to safeguard the forests and wildlife in the country" and "to protect and improve the natural environment including forests, lakes and rivers and wildlife, and to have compassion for the living creatures".

1.3 Nevertheless, over the years, there has been progressive pressure on the environment and the natural resources, the alarming consequences of which are becoming evident in increasing proportions. These consequences detract from the gains of development and worsen the standard of living of the poor who are directly dependent on natural resources. It is in this context that we need to give a new thrust towards conservation and sustainable development.

1.4 The National Conservation Strategy and the Policy Statement on Environment and Development are in response to the need for laying down the guidelines that will help to weave environmental considerations into the fabric of our national life and of our development process. It is an expression of our commitment for reorienting policies and action in unison with the environmental perspective.

2.0 Environmental Problem: Nature and Dimensions

2.1 Environmental problems in India can be classified into two broad categories:

- those arising as negative effects of the very process of development; and
- those arising from conditions of poverty and under-development.

The first category has to do with the impact of efforts to achieve rapid economic growth and development and continuing pressures of demand generated by those sections of society who are economically more advanced and impose great strains on the supply of natural resources. Poorly planned developmental projects are also often environmentally destructive. The second category has to do with the impact on the health and integrity of our natural resources (land, soil, water, forests, wildlife, etc.) as a result of poverty and the inadequate availability, for a large section of our population, of the means to fulfil basic human needs (food, fuel, shelter, employment, etc.). Needless to say, the two problems are interrelated.

2.2 Population is an important resource for development, yet it is a major source of environmental degradation when it exceeds the threshold limits of the support systems. Unless the relationship between the multiplying population and life support systems can be stabilized, development programmes, however, innovative, are not likely to yield the desired results. It is possible to expand the "carrying capacity' through technological advances and spatial distribution. But neither of these can support unlimited population growth. Although technological progress will add to the capabilities for sustaining a large number of population, the need for a vigorous drive for population control can hardly be over emphasized in view of the linkage between poverty, population growth and the environment.

2.3 Even today, over 250 million children, women and men suffer from under-nutrition. The scenario for the coming years is alarming and we are likely to face food crisis unless we are in a position to increase crop and animal productivity on a continuing basis, since the only option open to us for increasing production is productivity improvement. Also, access to food will have to be ensured through opportunities for productive employment.

2.4 A growth in domesticated animal population has been accompanied by a loss of area under grasslands and pastures. Hardly, 3.5 percent of our geographical area is under grasslands, while our domesticated animal population numbers nearly 500 million. The livelihood security of majority of our people depends on land and water based occupations such as crop and animal husbandry, forestry and fisheries.

2.5 Out of total area of India of about 329 million hectares, 175 million hectares of land require special treatment to restore such land to productive and profitable use. The degradation is caused by water and wind erosion (150 million hectares), salinity and alkalinity (8 million ha) and river action and other factors (7 million hectares).

2.6 Our forest wealth is dwindling due to over-grazing, over-exploitation both for commercial and house-hold needs, encroachments, unsustainable practices including certain practices of shifting cultivation and developmental activities such as roads, buildings, irrigation and power projects. The recorded forest cover in the country is 75.01 million hectares which works out to 19.5 percent of the total geographical area against the broad national goal of 33 percent for the plain areas and 66 percent for hilly regions. Even within this area, only 11 percent constitute forests with 40 percent or more of crown cover. According to the State of Forest Report, 1991, the actual forest cover in the country was 64.07 million hectares during 1987-89. The loss of habitat is leading to the extinction of plant, animal and microbial species. According to the Botanical and Zoological Surveys of India, over 1500 plant and animal species are in the, endangered category. The biological impoverishment of the country is a serious threat to sustainable advances in biological productivity. Gene erosion also erodes the prospects for deriving full economic and ecological benefits from recent advances in molecular biology and genetic engineering.

2.7 Our unique wetlands, rich in aquatic and bird life, providing food and shelter as also the breeding and spawning ground for the marine and fresh water fishes, are facing problems of pollution and over-exploitation. The major rivers of the country are also facing problems of pollution and siltation. Our long coastline is under similar stress. Our coastal areas have been severely damaged due to indiscriminate construction near the water-line. Coastal vegetation including mangroves and sea grasses is getting denuded. Our mountain ecosystems are under threat of serious degradation. Extensive deforestation leading to the erosion of valuable topsoil is threatening the livelihood security of millions of hill people. Equally serious is the downstream effects of the damage done upstream. Indo-gangetic agriculture, often described as a potential bread basket in the world, is being damaged beyond repair as a result of soil degradation. Some areas are facing problems of water-logging and rising water tables because of poorly planned and ill-executed irrigation. In other areas, the water table is receding because of over-exploitation of ground water. Furthermore, the quality of groundwater is being affected due to chemical pollution and in coastal areas, due to the ingress of sea water. The excessive use of fertilizers and pesticides impose threat to human health, to the genetic stocks and reduces the natural soil fertility in the long run. The absence of an integrated land and water use policy for the country is taking a heavy toll on these basic natural assets.

2.8 Coral reefs are the most productive marine ecosystems and provide habitat for diverse flora and fauna. These ecosystems are adversely affected by indiscriminate exploitation of coral for production of lime, recreational use and for ornamental trade. Similarly, the fragile environs of island ecosystems have been subjected to pressures of various forms including migration of people from the mainland.

2.9 Global atmospheric changes resulting in altered temperature and precipitation and rising ocean levels, are no longer within the realm of mere theoretical possibilities. Combination of local subsidence, greenhouse induced sea-level rise and coastal environmental degradation may lead to periodic floods, incursion of salt water, melting of glaciers and river flooding. Local changes of average rainfall will severely affect agriculture and water supply, especially in semi-arid areas.

2.10 Compounding these human-inflicted wounds on natural ecosystems and life-support mechanisms, we are facing serious problems of pollution and unsanitary conditions especially in urban areas. Pollution arising from toxic wastes and non-biodegradable consumer articles is tending to increase.

2.11 Lack of opportunities for gainful employment in

villages and the ecological stresses is leading to an ever increasing movement of resource-poor families to towns. Mega cities are emerging and urban slums are expanding. Illiteracy and child labour are persisting. There has been a substantial urban growth in the last four decades. This has resulted in congestion and squatter settlements with millions of people having no access to the basic needs of civic amenities. The green cover in our urban centres has been largely destroyed and once beautiful garden cities have become concrete jungles. The man-made heritage in India has been often gravely and even irrevocably damaged.

2.12 A large number of industries and other development projects have been incorrectly sited, leading, on the one hand, to over-congestion and over-pollution in our urban centres and on the other hand, to diversion of population and economic resources from the rural areas. This has also resulted in the pollution of most of our water bodies which are major constituents of our life support systems. Pollution of water bodies, in turn, has adversely affected the growth of aquatic fauna and flora which is an environmentally undesirable phenomenon for any ecosystem. The problems of women in villages are compounded in this whole scenario of energy, environmental and developmental imbalance. The incidence on malaria is high in many parts of the country. Safe drinking water is still a luxury in many villages. Liver ailments and gastro-intestinal diseases are common due to unclean drinking water.

2.13 It is difficult to clearly delineate the causes and consequences of environmental degradation in terms of simple one-to-one relationships. The causes and effects are often interwoven in complex webs of social, technological and environmental factors. For instance, from a purely scientific and technological standpoint, soil erosion would result from the cultivation of marginal lands. However, from the point of view of a comprehensive environmental impact analysis, it is important to go further back and analyze the circumstances that force people to cultivate marginal lands. Viewed in this light, it becomes clear that a concern for the environment is essentially a desire to see that national development proceeds along rational, sustainable lines. Environmental conservation is, in fact, the very basis of all development.

2.14 The overriding impact of adverse demographic pressures on our resources and ecosystems due to poverty and overpopulation of man and livestock has to be highlighted. Unless there is a curb on population growth and even a reduction of such populations and a corresponding improvement in land use policies, the current trend of over-exploitation and ecological degradation is not likely to improve.

2.15 Thus, we are faced with the heed for accelerating the pace of development for alleviation of poverty which is, to a great extent, responsible for many of our environmental problems. On the other hand, we have to avoid proceeding along paths with environmental costs so high that these activities cannot be sustained. Development has to be sustainable and all round, whether for the poor or the not so-poor or for the village folk or for the town people. The development models followed so far need to be reviewed.

3.0 Actions Taken

In recognition of the felt need for environmental protection, various regulatory and promotional measures have been taken in our country over the past twenty years. These include the following:

3.1 Legal

- The Wildlife (Protection) Act, 1972, amended in 1983, 1986 and 1991.
- The Water (Prevention and Control of Pollution) Act, 1974, amended in 1988.
- The Water (Prevention and Control of Pollution) Cess, Act, 1977, amended in 1991.
- The Forest (Conservation) Act, 1980, amended in 1988.

- The Air (Prevention and Control of Pollution) Act, 1981, amended in 1988.
- The Environment (Protection) Act, 1986.
- The Motor Vehicle Act, 1938, amended in 1988.
- The Public Liability Insurance Act, 1991.
- A Notification on Coastal Regulation Zone, 1991.

3.2 Institutions

- Department of Environment in 1980 and the integrated Ministry of Environment & Forests in 1985, Department of Science and Technology, Department of Agriculture and Cooperation, Department of Biotechnology, Department of Ocean Development, Department of Space, Department of Non-Conventional Energy Sources, Energy Management Centre, Council of Scientific and Industrial Research etc. at the Centre, Departments of Environment at the State and Union Territory level.
- Central Pollution Control Board and State Pollution Control Boards.
- Central Forestry Board.
- Indian Council of Forestry Research and Education with specialized institutions for research in and zone, forestry, moist and deciduous forests, wood technology, genetics and tree breeding and deciduous forests.
- Forest Survey of India (FSI) and the Wildlife Institute of India (VAI) in addition to the existing organizations like Botanical Survey of India (BSI) and Zoological Survey of India (ZSI).
- National Land-use and Wasteland Development Council.
- National Wastelands Development Board.
- Indian Board of Wildlife.
- National Museum of Natural History, Centre for Environmental Education, Institute for Himalayan Environment and Development and Centres of Excellence in specialized subject areas are among the various institutions set up.

3.3 Prevention and Control of Pollution

- Water and air quality monitoring stations in selected areas.
- Use-based zoning and classification of major rivers.
- Notification and enforcement of standards for polluting industries through the Central and State Pollution Control Boards.
- Rules for manufacture, storage, transportation and disposal of hazardous substances.
- On-site and off-site emergency plans for preparedness against chemical accidents.
- Fiscal incentives for installation of pollution control devices.
- Ganga Action Plan to prevent pollution of the river and restore its water quality which could be expanded to cover other major river systems subject to availability of resources.
- Identification of critically polluted areas and of highly polluting industries.

3.4 Conservation of Forests and Wildlife

- Adoption of a new Forest Policy (1988) with the principal aim of ensuring ecological balance through conservation of biological diversity, soil and water management, increase of tree cover, meeting the requirements of the rural and tribal population, increase in the productivity, efficient utilization of forest produce, substitution of wood and people's involvement for achieving these objectives.
- Under the Forest (Conservation) Act, 1980 stringent provisions for preventing diversion of forest land for any other purpose.
- Setting up of the National Wastelands Board to guide and oversee the wastelands development programme by adopting a mission approach for enlisting people's participation, harnessing the inputs of science and technology and achieving interdisciplinary coordination in programme planning and implementation.
- Formulation of a National Wildlife Action Plan.

- An exercise for preparation of a National Forestry Action Programme.
- Establishment of National Parks and Sanctuaries covering about 4 percent of the country's area.
- Eco-development plans for sanctuaries and National Parks.
- Identification of bio-geographical zones in the country for establishing a network of protected areas including seven Biosphere Reserves set up so far.
- Management Plans for identified wetlands, mangrove areas and coral reefs.
- Formulation of a National River Action Plan.

3.5 Land and Soil

- Surveys by the All India Soil, and Land-Use Survey Organization.
- Treatment of catchment in selected river valley projects and integrated watershed management projects in catchment of flood prone rivers.
- Assistance to States to control shifting cultivation.
- Assistance for reclamation and development of ravine areas.
- Drought prone areas programme.
- Desert development programme.

3.6 Environmental Impact Assessment

- Establishment of procedures for environmental impact assessment and clearance with regard to selected types of projects requiring approval of the Government of India.
- Prior clearance of projects requiring diversion of forests for non- forest purpose under the Forest (Conservation) Act, 1980.
- Formulation of Environmental guidelines for projects in various sectors.

3.7 Other Activities

- Eco-Task Forces of ex-servicemen for ecological restoration through afforestation and soil conservation.
- National Environmental Awareness Campaigns for

creating environmental awareness through non-governmental organizations.

- Surveys and research studies.
- Training programmes, workshops and seminars for building up professional competence and for creation of awareness.

4.0 Constraints and Agenda for Action

4.1 The modest gains made by the steps taken during the past few years leave no room for complacency when viewed in the context of enormous challenges. We can meet the challenges only by redirecting the thrust of our developmental process so that the basic needs of our people are fulfilled by making judicious and sustainable use of our natural resources. Conservation, which covers a wide range of concerns and activities, is the key element of the policy for sustainable development. Framing a conservation strategy is, therefore, an imperative first step. Development requires the use and modification of natural resources; conservation ensures the sustainability of development for the present and in the future. The conservation strategy is to serve as a management guide for integrating environmental concerns with developmental imperatives.

4.2 The primary purpose of the strategy and the policy statement is to include and reinforce our traditional ethos and to build up a conservation society living in harmony with Nature and making frugal and efficient use of resources guided by the best available scientific knowledge.

4.3 The agenda for action in this regard will include the following:

- to ensure sustainable and equitable use of resources for meeting the basic needs of the present and future generations without causing damage to the environment;
- to prevent and control future deterioration in land, water and air which constitute our life-support systems;
- to take steps for restoration of ecologically degraded areas

and for environmental improvement in our rural and urban settlements;

- to prevent further damage to and conserve natural and man-made heritage;
- to ensure that development projects are correctly sited so as to minimize their adverse environmental consequences;
- to ensure that the environment and productivity of coastal areas and marine ecosystems are protected;
- to conserve and nurture the biological diversity, genepool and other resources through environmentally sustainable development and management of ecosystems, with special emphasis on our mountain, marine and coastal, desert, wetlands, riverine and island ecosystems; and
- to protect the scenic landscapes, areas of geo-morphological significance, unique and representative biomes and ecosystems and wildlife habitats, heritage sites/structures and areas of cultural heritage importance.

4.4 To address to the above stated agenda, the instruments for action will include the following:

- to carry out environmental impact assessment of all development projects right from the planning stage and integrate it with their cost-benefit considerations. Appropriate costs of environmental safeguards and regeneration would continue to form an integral part of the projects;
- to ensure that all projects above a certain size and in certain ecologically sensitive areas should require compulsory prior environmental clearance;
- to incorporate environmental safeguards and protection measures, in policies, planning, site selection, choice of technology and implementation of development projects like agriculture, water resource development, industry, mineral extraction and processing, energy, forestry, transport and human settlements;
- to encourage research, development and adoption of environmentally compatible technologies; and to promote

application of the modern tools of science and technology for conservation, bridging of large gaps in supply and demand as well as control and monitoring of natural resources;

- to elicit and ensure participation of people in programmes for environmental improvement and for integrating the environmental concerns in planning and implementation of development programmes;
- to create environmental consciousness through education and mass awareness programmes;
- to aim at moderation of process of demand unleashed by the developmental process itself by taking measures to recycle waste materials and natural resources, conserve energy, conserve use of natural resources in industrial products by measures like wood substitution and generally try to reach moderation's in life styles consistent with sustainability and human dignity; to develop appropriate organizational structures and a pool of professional manpower to serve as the cadre for environmental management service; and
- to effectively implement the various environmental laws and regulations for environmental protection through creation or strengthening of the requisite enforcement machinery.

5.0 Priorities and Strategies for Action

5.1 Population Control

5.1.1 Unabated population growth, as at present, not only adds to the economic burden for all developmental activities, but also reduces the impact of economic growth on our society. Therefore, for the success of our planning, population control becomes the most urgent necessity. A comprehensive programme, with strong political backing and appropriate socio-economic measures, fully utilizing the available scientific know-how, simultaneously making efforts for developing new methodologies, and supported by modern communication technology and managerial and organizational skills, is essential for success in this most difficult area. Population control should be a national mission for the next decade. Despite efforts of several years, population control projects have not met with success. More stern measures such as legislative and better incentives are needed.

5.1.2 Along with the development programmes to improve the living conditions, action must be directed towards stabilization of population including the following measures:

- Launching a time bound national campaign for population stabilization with the small family as a socially responsible objective;
- Increased support for female education, female employment, and of social security programmes;
- Easier access to the means of family planning and health care facilities;
- Added incentives in terms of taxation and other benefits for family planning;
- Environmental sanitation, prevention and control of communicable diseases through integrated vector control and health education; and
- Adoption of decentralized renewable energy devices that enhance quality of life in remote pockets while taking special care of the health needs of women.

5.2 Conservation of Natural Resources

5.2.1 Land and Water

5.2.1.1 An integrated land and water management approach is extremely important to sustain the food production, animal husbandry and other activities.

5.2.1.2 Amelioration of water-logged and salt affected lands, command area development, protection of good agricultural land against diversion to urban and other uses, prevention of land fragmentation, maintenance of sustained productivity of soil and conservation of lands with forests and vegetal cover are the integral components of sustainable management.

5.2.1.3 The importance of water as a finite, though a renewable resource, must be clearly recognized. Land and water use are to be considered together, particularly in the context of recurring droughts and floods. Water conservation measures; discipline on use of water; economizing the consumption of water in households, agriculture and industry; and appropriate recycling would be essential.

5.2.1.4 The steps to be taken for sustainable use of land and water should include the following:

- Classification, zoning and apportionment of land for designated uses such as, agriculture, forestry, grassland, green areas, industrial activities, catchment areas and watersheds and human settlements based on assessment of their capabilities and environmental considerations;
- Enactment of laws for appropriate land uses to protect the soil from erosion, pollution and degradation;
- Protection of land near water bodies and prevention of construction there upon;
- Measures to ensure equitable access to and responsibility for sustainable use of land and water resources;
- Micro-level planning to develop appropriate methodology and implementation of action plan by involving the people at the village level in social forestry programmes, land use planning, afforestation etc.;
- Countrywide campaign to minimize soil and run-off losses by carrying out extensive works like contour trenching, contour bounding, terracing, construction of small storages, catchment treatment and protection of the vegetal cover in the catchments and watersheds. This is to be a specific charge of project authorities in all irrigation, power, road and agricultural projects;
- Restoration and reclamation of degraded areas including weed infested areas, mined areas, grazing lands and salt affected soils;
- Measures for preventing wind erosion by undertaking special programmes of conservation and afforestation in

desert areas;

- Development of suitable agro-silvipastoral techniques with special emphasis on hilly areas and in, and semi-arid zones;
- Building up a network for assessment and monitoring of soil and water (surface and ground water) quality throughout, the country which should be on a permanent basis as in the case of meteorological stations;
- Measures for water conservation, recycling and optimal conjunctive use of surface and ground water for specific uses;
- Legislative measures to check over-exploitation of surface and ground water for various uses;
- Conservation of wetlands for ensuring sustainable ecological and economic benefits;
- Encouragement to and improvement in traditional methods of rain water harvesting and storage;
- Stringent measures for prevention and control of pollution due to indiscriminate disposal of solid wastes,, effluents and hazardous substances in land and water courses;
- Control and abatement of pollution of water bodies from municipal and industrial wastes generated from urban habitats by intercepting and diverting such wastes away from water bodies;
- Classification, zoning and regulations for maintaining the quality of the water bodies to protect and enhance their capabilities to support the various designated uses; and
- Adoption of low cost sanitation technology for prevention and control of pollution in water courses.

5.2.2 Atmosphere

5.2.2.1 For prevention and control of atmospheric pollution including noise pollution, the thrust will be on the following:

- Use of clean fuels and clean technologies, energy efficient devices and air and noise pollution control systems;
- Setting up of source specific and area wise air quality

standards and time bound plans to prevent and control pollution;

- Proper location of projects to minimize the adverse impact on people and environment;
- Incentives for environmentally benign substitutes, technologies and energy conservation;
- Raising of green belts with pollution tolerant species;
- Developing coping mechanisms for future climatic changes as a result of increased emission of carbon dioxide and greenhouse gases; and
- Appropriate action to control adverse impact on Indian continent due to ozone depletion and other gaseous effects in the atmosphere at global level.

5.2.3 Biodiversity

5.2.3.1 About 90 percent of the world food comes from 20 plant species. The plant breeders find that they have to turn more and more to the wild species to introduce into the cultivated forms desired qualities of resistance to pests and diseases and the ability to withstand adverse soil and weather conditions. India's biological diversity is very rich but unfortunately its wealth is being eroded due to various reasons. This diversity needs to be preserved and the immediate task will be to devise and enforce time bound plans for saving the endangered plant and animal species as well as habitats of biological resources. Action for conservation must be directed to:

- Intensification of surveys and inventorisation of biological resources in different parts of the country including the island ecosystems. The survey should include information on distribution pattern of particular species, population and communities and the status of ethnobiologically important groups;
- Conservation of biodiversity through a network of protected areas including Biosphere Reserves, Marine Reserves, National Parks, Sanctuaries, Gene Conservation Centres, Wetlands, Coral Reefs and such other natural

habitats of biodiversity. This should include taxonomic and ecological studies on the flora and fauna with adequate emphasis placed on the lower vertebrate, invertebrate and micro-flora which are important in contributing to the healthy maintenance of ecosystems;

- Full and correct rehabilitation of rural poor/tribals displaced due to creation of national parks/biosphere reserves/tiger reserves;
- Conservation of micro-fauna and micro-flora which help in reclamation of wastelands and revival of biological potential of the land;
- Protection and sustainable use of plant and animal genetic resources through appropriate laws and practices;
- Protection of domesticated species/varieties of plants and animals in order to conserve indigenous genetic diversity;
- Maintenance of corridors between national parks, sanctuaries, forests and other protected areas;
- Emulation and support for protecting traditional skills and knowledge for conservation;
- Development of methodologies to multiply, breed and conserve the threatened and endangered species through modern techniques of tissue culture and biotechnology;
- Discouragement of monoculture and plantation of dominating and exotic species, in areas unsuited for them and without sufficient experimentation; and
- Restriction on introduction of exotic species of animals without adequate investigations.

5.2.4 Biomass

5.2.4.1 For the vast majority of our rural people, the foremost need is for fuel wood, timber, fodder, fibre, etc. The issue of sustainable resource utilization, therefore, has to be specially addressed first from the point of view of the biomass requirements of the rural poor. Action must be directed to:

- Devising ways and means by which local people can conserve and use thereafter the resources of the common lands and degraded forests, so that they have a stake in the

continuing productivity of the resources;
- Encouraging private individuals and institutions to regenerate and develop their wastelands;
- Raising of fuel-wood species and provision of alternatives to reduce dependence on fuel-wood;
- Taking measures to increase the production of fodder and grasses to bridge the wide gap between supply and demand;
- Raising of bamboo and species providing small timber for local house-construction and agricultural implements;
- Increasing biomass to meet essential requirement of biomass based industry;
- Promoting direct relationship between forest-based industry and farmers to raise needed raw materials, provided this does not result in diversion of prime agricultural lands and displacement of small and marginal farmers;
- Extensive research and development in forestry for better regeneration and improved productivity;
- Development of technologies for enhancing the productivity and efficiency of use of all biomass resources (both terrestrial and marine);
- Institutional and technological systems to enable rural artisans to sustain biomass based crafts; and
- Curtailment of the supply of subsidized biomass based resources to industrial consumers.

6.0 Development Policies from Environmental Perspectives

Implementation of the aims and objectives of conservation and sustainable development will require integration and internalization of environmental considerations in the policies and programmes of development in various sectors.

Curtailment of consumerism and shift towards use of environment friendly products and processes, and low waste generating technologies through conscious efforts and appropriate economic policies including pricing of natural resources as well as fiscal incentives and disincentives will be the guiding factors for ensuring conservation and sustainable development.

For environmental conservation and sustainable development, the steps which need to be taken in some of the key sectors of development activities are outlined in the following sections.

6.1 Agriculture and Irrigation

For sustainable management of agriculture and irrigation, the action points should include the following:

6.1.1 Agriculture

- Development of pesticides and insecticides policy for the country;
- Development of integrated pest management and nutrient supply system;
- Development and promotion of methods of sustainable farming, especially organic and natural farming;
- Efficient use of inputs including agro-chemicals with minimal degradation of environment;
- Phasing out and stoppage of persistent and toxic pesticides and their substitution by environmentally safe and appropriate pesticides;
- Promotion of environmentally compatible cropping practices, bio-fertilisers and bio-pesticides;
- Restriction on diversion of prime agricultural land for other purposes;
- Ensuring land for different uses based upon land capability and land productivity;
- Evolving cost effective and efficient methods of water conservation and use;
- Incentives for cultivation of crops with high nutritive value and those with lesser demands on water and energy inputs;
- Encouraging crop rotation patterns;
- Strengthening of local bodies like Zilla Parishads, Panchayats and Samitis to ensure effective decentralization and optimal resource management; and

- Anticipatory programmes and contingency plans for disasters such as drought, flood and climate change.

6.1.2 Irrigation

- Priority to small projects to meet the requirements of irrigation without causing significant alteration in the environmental conditions;
- Revival of traditional water management systems and development of alternate irrigation systems such as harvesting and conservation of run-off rain water;
- Measures for increasing the efficiency of water-use, water conservation and recycling;
- Measures for provision of drainage as an integral component of irrigation projects and to prevent water logging and leaching;
- Watershed management through catchment treatment of the drainage areas, protection of vegetal cover and measures to prevent siltation in an integrated manner with the irrigation authorities being fully accountable; catchment treatment would be so designed as to have a direct impact on the life of the reservoir, hydrological regime and life support systems. It would depend on the location specific conditions in each case;
- Adoption of command area development approach for all irrigation projects to ensure optimal utilization;
- Critical assessment of irrigation projects and delivery systems to ensure optimal utilization of water resources along with measures to mitigate environmental and social damage;
- Focus on decentralized network of small irrigation and water projects with minimum environmental disruption which will be of great value to local communities and yet capable of generating surplus for other areas at low cost;
- Design and implementation of irrigation projects which are environmentally sustainable, based on lessons learnt from a critical analysis of all past projects; and
- Continuous and ongoing evaluation and monitoring of all

projects.

6.2 Animal Husbandry

The activities relating to animal husbandry should concentrate on the following:

- Development of an animal husbandry policy for the country;
- Intensification of sterilization programme for containing unsustainable growth in livestock population;
- Improvement in genetic variability of indigenous population;
- Distribution of animals like goats under the Integrated Rural Development Programme strictly consistent with the availability of pasture lands to reduce pressure on the lands;
- Propagation of wildlife and wildlife resources management on sustainable basis;
- Selective breeding of animals used for draught power to conserve fuel;
- Promotion of stall feeding and rotational grazing;
- Restoration and protection of grazing lands;
- Involvement of local people in the policy planning on pasture lands and stall feeding to avoid fodder scarcity; and
- Incentive for growing fodder crops and establishment of fodder banks.

6.3 Forestry

Concerted efforts should be made for raising the forest cover and for conservation of existing forests which constitute an essential life support system and an important source of food, fibre, fodder, fuel and medicines etc. For attaining the goal of having at least one third of our land area under forest cover, intensified measures on a mission mode are required to be taken along with commensurate mobilization of resources for this purpose. As outlined in the National Forest Policy (1988), the action points should include the following:

- Maintenance of environmental stability through

preservation and, where necessary, restoration of the ecological balance that has been adversely disturbed by serious depletion of the forests of the country;

- Conserving the natural heritage of the country by preserving the remaining natural forests with the vast majority of flora and fauna, which represent the biological diversity and genetic resources of the country;
- increasing substantially the forest/tree cover in the country through massive afforestation and social forestry programmes, especially on all denuded, degraded and unproductive lands involving the local people in this endeavour by giving them tangible economic motives and employment opportunities;
- Meeting the rights and concessions for requirements of fuel wood, fodder, minor forest produce and small timber of the rural and tribal population with due cognizance of the carrying capacity of forests;
- Increasing the productivity of forests to meet the essential national needs;
- Encouraging efficient utilization of forest produce;
- Restriction on diversion of forest lands for non-forest uses and compensatory afforestation in case where diversion is unavoidable;
- Afforestation on common lands by the local communities through usufruct-sharing schemes;
- Motivation of farmers/land owners to resort to tree farming in similar manner of crop based farming;
- Substitution of wood by other materials, alternative sources of energy and fuel efficient stoves;
- Permission to forest-based enterprises after a thorough scrutiny regarding the availability of raw materials;
- Supply of forest produce to the industrial consumers only at its true market value and not at concessional prices;
- Involvement of local people and dedicated grass roots nongovernmental organizations, in the afforestation programme and for protection of existing forests; and

- Creation of land banks for compensatory afforestation.

6.4 Energy Generation and Use

For prevention and control of pollution and environmental hazards in energy generation and use as also for encouraging popularization of environmentally benign energy systems, the following measures should be taken:

- Environmental impact assessment prior to investment decisions and site selection; choice of practicable clean technologies for energy production and processes including waste utilization, treatment and disposal of solid wastes, effluents and emissions;
- Location of energy generation projects based on environmental considerations including pollution, displacement of people and loss of biodiversity;
- Decentralized small projects for meeting the rural energy needs and incentives for use of non-conventional energy sources;
- Incentives and punitive measures (including proper pricing) to prevent abuse and to promote the use of energy efficient devices in the production and distribution systems and for energy conservation in all sectors including households, agriculture, industry, power and transportation;
- Concerted efforts for development and propagation of non-conventional renewable energy generation systems; and
- Setting up of biogas plants based on cow-dung, human excreta and vegetable wastes.

6.5 Industrial Development

Environmental considerations should be integrated while encouraging industrial growth. The action points in this regard should include a mix of promotional and regulatory steps which are as follows:

- Incentives for environmentally clean technologies, recycling and reuse of wastes and conservation of natural resources;
- Operationalisation of 'polluter pays principle' by

introducing effluent tax, resource cess for industry and implementation of standards based on resource consumption and production capacity;

- Fiscal incentives to small-scale industries for pollution control and for reduction of wastes;
- While deciding upon sites, priority to compatible industries so that, to the extent possible, wastes from one could be used as raw material for the other and thus the net pollution load is minimized;
- Location of industries as per environmental guidelines for sitting of industry;
- Enforcement of pollution control norms in various types of industrial units depending on their production processes, technologies and pollution potential; particular attention to be paid to highly polluting industries;
- Encouragement for use of environmentally benign automobiles/motor vehicles and reduction of auto-emissions;
- Collective efforts for installation and operation of common effluent treatment facilities in industrial estates and in areas with a cluster of industries;
- Introduction of 'Environmental Audit' and reports thereof to focus on environment related policies, operations and activities in industrial concerns with specific reference to pollution control and waste management;
- Dissemination of information for public awareness on environmental safety aspects and stringent measures to ensure safety of workers and general population against hazardous substances and processes;
- Preparation of on-site emergency plans for hazardous industries and off-site emergency plans for districts in which hazardous units are located;
- Public liability insurance against loss or injury to life or property;
- Setting up of environment cells in industries for implementing environmental management plans and for

compliance of the requisites of environmental laws;

- Internalizing the environmental safeguards as integral component of the total project cost;
- Environmental impact assessment from the planning stage and selection of sites for location of industries; and
- Clearance by Ministry of Environment and Forests of all projects above a certain size and in certain fragile areas.

6.6 Mining and Quarrying

To prevent and to mitigate environmental repercussions in mining and quarrying operations, action must be directed to:

- Mined area rehabilitation and implementation of the environmental management plans concurrently with the on-going mining operations to ensure adequate ecological restoration of the affected areas;
- Rehabilitation of the abandoned mined areas in a phased manner so that scarce land resources can be brought back under productive use;
- Laying down of requisite stipulations for mining leases regarding tenure, size, shape and disposition with reference to geological boundaries and other mining conditions to ensure systematic extraction of minerals along with environmental conservation;
- Emphasis on production of value added finished products from mining so as to reduce indiscriminate extraction;
- Upgradation and beneficiation of minerals at the source, to the extent possible in order to ensure utilization of low-grade mineral resources and to reduce the cost of transportation, processing and utilization;
- Environmentally safe disposal of the by-products of mining;
- Restriction on mining and quarrying activities in sensitive areas such as hill slopes, areas of natural springs and areas rich in biological diversity;
- Discouraging selective mining of high grade ores and recovery of associated lower grade ores during mining; and,

- Environmental impact assessment prior to selection of sites for mining and quarrying activities.

6.7 Tourism

To ensure sustainable growth of tourism without causing irreversible damage to the natural environment, activities relating to tourism should take care of the following:

- Promotion of tourism based on careful assessment of the carrying capacity and support facilities such as transport, fuel, water and sanitation;
- Development of tourism in harmony with the environmental conditions and without affecting the lifestyles of local people; and,
- Restriction on indiscriminate growth of tourism and strict regulation of the tourist activities in sensitive areas such as hill slopes, islands, coastal stretches, National Parks and Sanctuaries.

6.8 Transportation

For prevention of pollution and for development of environmentally compatible transportation systems, the following steps should be taken:

- Improvement in mass transport system to reduce increasing consumption of fuel, traffic congestion and pollution;
- Improved transport system based on bio-energy and other non-polluting energy sources;
- Rail transport and pipeline transport instead of road transport, where ever possible, by appropriate freight pricing so as to reduce congestion, fuel consumption and environmental hazards;
- Transportation of hazardous substances through pipelines;
- Improvement in traffic flow through proper maintenance of roads, updated traffic regulation and strict enforcement of prescribed standards;
- Enforcement of smoke emission standards for containing vehicular exhausts, at the manufacturer and user level;
- Phasing out the use of lead in motor spirit; and

- Regulations for environmental safety in transportation of hazardous substances.

6.9 Human Settlements

To check unplanned growth of human settlements and to ensure a better quality of life for the rural and urban population, the action points should include the following:

- Creation of gainful employment opportunities and provision for meeting the basic needs through better communications, entertainment, medical and educational facilities in rural areas to check rural-urban migration;
- Decentralization of urbanization through establishment of secondary cities and towns with requisite infra-structural services and employment opportunities by developing human settlement perspective plan at national and state level;
- Disincentives for industrial and job location in existing urban centres which have exceeded their carrying capacity;
- Improvement of infrastructural facilities such as water supply, sewerage, solid waste disposal, energy recovery systems and transportation in an integrated manner;
- Promoting the use of indigenous building materials and appropriate construction technologies by revising building and planning codes supporting small scale production, skill upgradation of artisans and people oriented delivery systems;
- Conservation of heritage sites and buildings, through regulation to ensure that these are not demolished, encroached upon and affected by indiscriminate construction and pollution;
- Stock-taking of buildings, areas, monuments of heritage value in the country;
- Recycling of existing building stock to save green open compounds and save building material;
- Planning of shade giving and fruit bearing and ornamental trees along the road side, in the compounds of schools, hospitals, Government as well as private office buildings,

places of worship, places meant for public fairs, assemblies and markets, and the periphery of play grounds and water bodies;
- Botanical gardens representing the local flora;
- Raising of gardens, parks and open spaces in the towns and cities for public use and for promotion of environmental consciousness;
- Laying down a system for the propagation and protection of urban forestry by assigning responsibility amongst the various authorities;
- Deterrent measures to discourage indiscriminate growth of human settlement and polluting industries in vulnerable areas such as hilly regions and coastal stretches;
- Environmental appraisal of projects related to urban development and regional planning, preparation of environmental/eco-development plans for sensitive regions and sub-regions for evolving desirable norms and space standards;
- Prevention of environmental health problems and associated communicable and non-communicable diseases by educating people on personal hygiene, sanitation and use of potable water;
- Creation/strengthening of health care facilities for all sections of society both in rural and urban areas; and
- Establishment of monitoring systems and epidemiological data to ensure adequate early warning system for prevention and control of diseases.

7.0 International Co-operation

7.1 A major threat to sustainable development has been visualized, in recent years, from environmental problems of a global nature-ozone layer depletion, global warming and climate change, destruction of biological diversity, trans-boundary air pollution, marine pollution and land-based marine pollution, trans-boundary movement of hazardous 'substances. On a philosophical plain, the scientific proof of such problems provides an opportunity to reconsider the development path ushered in by the industrial revolution, and the blinkered pursuit of lifestyles which place extreme pressures on the natural resource base. But at a practical level, it means pressures on developing countries to take measures which they can ill-afford.

7.2 It has been India's firm conviction that it is the process of industrialization, and the continued profligacy of industrialized economies that have created the problems which threaten our planet and its life forms. Not only do they use up non-renewable natural resources in disproportionate quantities, but create discharges and emissions which disturb delicate balances in eco-systems and atmospheric equilibrium. It is true, of course, that this has not been done consciously or intentionally (except in matters such as dumping of hazardous wastes, or the use of nuclear and chemical weapons). Nevertheless, the responsibility is clearly established, as also the need for urgent and effective action, by the developed world, to prevent global disaster. This includes not only direct action, but also indirect measures such as creation of an economic order which helps developing countries to exert less pressure on their own natural resources.

7.3 The Indian approach to global environmental problems is generally in keeping with other developing countries and has the following basic elements:

- Our economic development cannot be hampered in the name of the global environment, which we have done nothing to damage and can do little to save. Our resources are required to meet our developmental needs such as education, nutrition, health services, drinking water, housing, sanitation, agriculture, industry, infrastructure, even all of which we find it difficult to provide having been behind in the race for development. Without this development, threats to the environment will in any case grow. In the short run, this developmental effort could even add to the discharges and emissions which cause global problems, but these are miniscule compared to the quantities which industrialized countries have already contributed. In any case, such emissions etc., can easily be compensated for a marginal reduction of the same in the industrialized world;

- For environmental protection and improvement, we will do our best with the resources available in the country. With new and additional funding support and transfer of environmentally sound technologies from the developed countries, we will be in a position to augment our capacity to deal with the environmental problems; and
- Regulatory international regimes can be useful in some areas such as ozone depletion or even climate change, provided the special situation of developing countries is fully addressed. But in other sectors - such as forestry - such a regime is neither workable nor acceptable. In such sectors, what is required is a reduction of international economic and commercial pressures which generate unsustainable exploitation, and additional financial resources to tackle the damage already done.

7.4 India's traditional lifestyle still followed by a vast majority of its population has always emphasized conservation of plant and animal life, waste minimization, recycling, simplicity in food habits and other such environment - friendly attitudes. There is no doubt that with economic development will also come lifestyles which require more intensive resource use. On the one hand we have to minimize the adverse environmental impacts of development (e.g., through legislation and control, impact assessment and monitoring, education and awareness). On the other, we have to continuously see how far the traditional Indian ethos can be reflected in modern lifestyles. If these efforts are supported by adequate financial resources from the international community, as well as the transfer (and development) of environmentally sound technology, India can contribute significantly to the international action to deal with global environmental problems.

8.0 Support Policies and Systems

Implementation of the aims and objectives of environmental policy will need support policies and systems for filling up of the gaps in the existing institutional set up, legislative instruments and enforcement mechanisms, research and development, mobilization of financial resources, creation of public awareness and training of professionals.

8.1 Strengthening of Institutions and Legislation

8.1.1 It will require strengthening of existing institutions at different levels. It will need a close linkage among the compartmentalized sectors which have been historically dealt with by separate organizations. It will call for a change in the institutional mechanism for enlisting public participation. It will necessitate quick decision making on development projects based on assessment of their potential of rendering long term sustainable benefits to the society at large, particularly vulnerable sections.

It will also require effective implementation of laws and regulations for environmental protection through strengthening of and closer interaction among the regulatory bodies and administrative machinery.

8.1.2 Existing laws and enforcement mechanisms should be subjected to periodic review to evaluate their adequacy and efficacy in the light of changed circumstances and experience.

8.2 Natural Resource Accounting

8.2.1 As economic policies form the frame-work for a range of sectoral development, it will be necessary to consider how these policies affect the quality and productivity of environmental resources. This will require a system of resource accounting along with the other exercises of cost benefit analyses.

8.2.2 In essence, indicators of growth such as GNP and GDP should include a measure of depletion cost and value judgments in terms of environmental resources. It will require instruments and expertise for evaluation and conscious trade offs, where unavoidable, to meet the legitimate development needs.

8.2.3 The Government will prepare, each year, a natural resources budget which will reflect the state and availability of resources like land, forests, water etc. and which will rationally allocate these resources in keeping with the principles of conservation and sustainable development.

8.3 Training and Orientation Programmes

8.3.1 Available management resources in the enterprises/projects would be oriented towards environmental considerations and expertise to be developed through appropriate training programmes.

8.3.2 Formal education and training programme in specialized areas of pollution control and environmental management will be a continuing need. For this purpose, intensive programmes for education and training will need to be introduced in the universities, IITs and other professional institutions. Environmental education at the school level including training of teachers shall be an important component of educational programmes.

8.4 Promoting Environmental Awareness

To raise public awareness and involvement in environmental activities, the mass media ranging from local folklores to electronic media should serve as a vital role. To raise public awareness on environmental issues and to promote people's participation, in environmental activities and conservation of natural resources, development of environmental education resource material and use of traditional and modem media of communication need to be strengthened. Scope and functions of the existing environmental education centres should be further strengthened and enlarged to develop a network of infrastructure for environmental education including development of orientation centres and provision of educational material for visitors at the national parks, sanctuaries and tiger reserves.

8.5 Promoting Appropriate Environmental Technologies

Existing research and development efforts need to be strengthened to develop the appropriate low cost technologies considering the possibilities, opened up by biotechnology, genetic engineering, information and material technologies and remote sensing, tailored to the local environmental conditions.

8.6 Rehabilitation of Project Oustees

8.6.1 While implementing the projects in various sectors, conscious efforts should be made to avoid displacement of local people. Where it is unavoidable, necessary measures should be taken to ensure their rehabilitation by providing suitable facilities.

8.6.2 The Government will formulate a comprehensive national rehabilitation policy which, apart from other things, ensures that the oustees are economically better off than before and above poverty line as a result of their rehabilitation.

8.7 Role of Non-Governmental Organizations

8.7.1 Implementation of the conservation strategy would be impossible without active participation of the people. Non-governmental Organizations (NGOs) can play an important role in mobilizing the people at grassroots. This will need a network among NGOs and interface between people and Government to work on community involvement, providing information on environmental surveillance and monitoring, transmitting development in science and appropriate technology to the people at large.

8.7.2 Environmental Information Centres should be set up at the district level to generate knowledge regarding traditional and endogenous system management practices. NGOs at the district level should be involved in the management and dissemination of the environmental information.

8.7.3 Non-Governmental Organizations, citizen groups and village level institutions like forests panchayats, and Gram Sabhas should be empowered with locus standi and. support for mobilization of public opinion and participation in development activities.

8.7.4 Managerial capacity of the NGOs should be strengthened. Training programmes for NGOs on regional basis should be organized. An advisory cell for rural NGOs should be made available at all district headquarters.

8.8 Women and Environment

Women at the grassroots level should be actively involved in the conservation programmes which should be income generating and self financing and sustainable on a long term basis and the Government Ministries/Departments should have NGOs cell or at least a Liaison Officer for interaction with the NGOs.

8.9 Partnership Role of Centre and State Governments

Effective implementation of necessary measures, as outlined in the Statement, will be facilitated by a partnership role of Central and State Governments. Many environmental problems assume national significance. Hence, the policies and programmes at the State and Central level should be drawn up keeping in view overall national policy considerations. A monitoring mechanism involving central and state Government representatives will be set up for inter-action as required for implementation of the policy initiatives.

9.0 Conclusion

9.1 It is only through such initiatives the contours of which have been highlighted in preceding paragraphs, we will be in a position to resolve the conflicts which often arise between the environmental concerns and developmental pursuits that have a direct bearing on the very fabric of our society and life styles.

9.2 The task before us would be daunting if it were not for the many positive factors that are emerging: people's movements to conserve their own environment, greater public and media concern for environmental issues and spread of environmental awareness among children and youth.

9.3 It is up to us, as State and citizens, to undertake development process in keeping with our heritage and the traditional conservation ethos and in harmony with the environmental imperatives of this land.

Index

Index